Cell and Developmental Biology
of the Eye

Cell and Developmental Biology
of the Eye

Series Editors
Joel B. Sheffield and S. Robert Hilfer

Ocular Size and Shape: Regulation During Development

Cellular Communication During Ocular Development

Molecular and Cellular Basis of Visual Acuity

Heredity and Visual Development

Development and Order in the Visual System

The Microenvironment and Vision

Cell Interactions in Visual Development

The Proceedings of the Philadelphia Symposia on Ocular
and Visual Development

S. Robert Hilfer
Joel B. Sheffield
Editors

Cell Interactions in
Visual Development

Springer-Verlag
New York Berlin Heidelberg
London Paris Tokyo

S. Robert Hilfer
Joel B. Sheffield
Department of Biology
Temple University
Philadelphia, Pennsylvania 19122, U.S.A.

Library of Congress Cataloging-in-Publication Data
Cell interactions in visual development.
  (Cell and developmental biology of the eye)
  Based on the 11th Symposium of Ocular and Visual
Development, held Oct. 1987 at Temple University.
  Bibliography: p.
  Includes index.
  1. Eyes—Differentiation—Congresses.   2. Eyes—
Growth—Congresses.   3. Eyes—Cytology—Congresses.
I. Hilfer, S. Robert.   II. Sheffield, Joel B.
III. Symposium on Ocular and Visual Development
(11th : 1987 : Temple University)   IV. Series.
[DNLM: 1. Cell Communication—congresses.   2. Eye—
cytology—congresses.   3. Eye—growth & development—
congresses.   4. Vision—physiology—congresses.
WW 101 C3935   1987]
QP475.C47   1988   599'.01823   88-24757
ISBN 0-387-96829-6

QP
475
.C47
1987

Camera-ready text prepared by the editors.
Printed and bound by Arcata Graphics/Halliday, West Hanover, Massachusetts.
Printed in the United States of America.

9 8 7 6 5 4 3 2 1

ISBN 0-387-96829-6 Springer-Verlag New York Berlin Heidelberg
ISBN 3-540-96829-6 Springer-Verlag Berlin Heidelberg New York

Series Preface

The eye has fascinated scientists from the earliest days of biological investigation. The diversity of its parts and the precision of their interaction make it a favorite model system for a variety of developmental studies. The eye is a particularly valuable experimental system not only because its tissues provide examples of fundamental processes, but also because it is a prominent and easily accessible structure at very early embryonic ages.

In order to provide an open forum for investigators working on all aspects of ocular development, a series of symposia on ocular and visual development was initiated in 1973. A major objective of the symposia has been to foster communication between the basic research worker and the clinical community. It is our feeling that much can be learned on both sides from this interaction. The idea for an informal meeting allowing maximum exchange of ideas originated with Dr. Leon Candeub, who supplied the necessary driving force that made the series a reality. Each symposium has concentrated on a different aspect of ocular development. Speakers have been selected to approach related topics from different perspectives.

This book series, "Cell and Developmental Biology of the Eye," is derived from the Philadelphia symposia on ocular and visual development. Previous volumes are listed on the series page. We hope that the introduction of this proceedings series will make the results of research on ocular cell and developmental biology more widely known and more easily accessible.

Preface

The 11th Symposium on Ocular and Visual Development was held in October, 1987 to review the current understanding of cellular interactions in the development of the visual system. This volume is derived from the presentations at that symposium. Over the past few years, advances in monoclonal antibody technology and cell surface biochemistry have revealed a complex set of cell surface molecules that appear to convey information between different cells of the embryonic retina and control the ultimate organization of the tissue. Many of the chapters in this volume are concerned with these molecules and how they interact with other factors to create the ordered structure of the mature system. These studies have been aided by the development of sensitive techniques for marking cells so that their fates can be traced. Other chapters discuss the role of the Müller cells in retinal organization. Increasing evidence suggests that interactions between Müller cells and neuronal cells direct the development and maturation of the tissue. This has become particularly significant in view of recent studies which indicate that proliferative vitreoretinopathy, a major cause of blindness after retinal reattachment surgery, may arise through inappropriate stimulation of division in Müller cells.

We are grateful to Dr. E. Gruberg of the Biology Department at Temple for his expertise and assistance in the organization of the meeting, to the speakers for their presentations and contributions to this volume, and to the reviewers of the manuscripts for their helpful comments.

This symposium could not have been held without the generous support of the Provost and the College of Arts and Sciences of Temple University. We also thank the Albert B. Millett foundation for additional funding. This volume was prepared with the skillful assistance of George Basila, to whom we are indebted.

S. Robert Hilfer
Joel B. Sheffield

Philadelphia, Pennsylvania

Contents

Contributors

Leo M. Chalupa, Department of Psychology and the Physiology Graduate Group, University of California, Davis, California 95616, U.S.A.

Kathryn Farr, Department of Neurobiology, Anatomy, and Cell Science and The Center for Neuroscience, University of Pittsburgh, Pittsburgh, Pennsylvania 15261, U.S.A.

Scott E. Fraser, Department of Physiology and Biophysiology, University of California, Irvine, California 92717, U.S.A.

Deborah E. Hall, Howard Hughes Medical Institute, University of California, San Francisco, California 94143-0724, U.S.A.

Robert E. Hausman, Department of Biology, Boston University, Boston, Massachusetts 02215, U.S.A.

Carl Lagenaur, Department of Neurobiology, Anatomy, and Cell Science and The Center for Neuroscience, University of Pittsburgh, Pittsburgh, Pennsylvania 15261, U.S.A.

Vance Lemmon, Department of Neurobiology, Anatomy, and Cell Science and The Center for Neuroscience, University of Pittsburgh, Pittsburgh, Pennsylvania 15261, U.S.A.

Paul J. Linser, The C.V. Whitney Laboratory and the Department of Anatomy and Cell Biology, University of Florida, St. Augustine, Florida 32086, U.S.A.

Ronald L. Meyer, Department of Developmental and Cell Biology, University of California, Irvine, California 92717, U.S.A.

Karla M. Neugebauer, Howard Hughes Medical Institute, University of California, San Francisco, California 94143-0724, U.S.A.

Nancy A. O'Rourke, Department of Physiology and Biophysiology, University of California, Irvine, California 92717, U.S.A.

Pamela A. Raymond, Department of Anatomy and Cell Biology and The Neuroscience Program, University of Michigan, Ann Arbor, Michigan 48109-0616, U.S.A

Louis F. Reichardt, Howard Hughes Medical Institute, University of California, San Francisco, California 94143-0724, U.S.A.

P. Vijay Sarthy, Departments of Ophthalmology, Physiology, and Biophysics, University of Washington, Seattle, Washington 98195, U.S.A.

Photoreceptor–Müller Cell Interactions: Effects of Photoreceptor Degeneration on GFAP Expression in Müller Cells
P. Vijay Sarthy

Most neurons in vertebrate and invertebrate nervous systems are surrounded by satellite cells know as glia. From observations based primarily on neuropathological cases, anatomists at the beginning of the century, such as Virchow, Golgi and Cajal, proposed a neuron-glia relationship in which glial cells were involved in providing mechanical support, nutrients, and electrical insulation to neurons. Glia have now been implicated in diverse processes such as neurotransmitter inactivation, K^+-homeostasis, and the migration and differentiation of neurons (Somjen and Varon, 1979; Purves and Lichtman, 1986).

Although the role of glia in neuronal development and metabolism has received considerable attention, the influence of neurons on glia is less well studied. Developmental studies show that neurons can modulate the expression of certain glial proteins. For example, Schwann cell differentiation and high level synthesis of myelin proteins depends on contact with axons (Bentley et al., 1981; Lees and Brostoff, 1981). Similarly, the secretion of basal lamina components such as type IV collagen by Schwann cells is dependent on the presence of neurons (Bunge et al., 1986). In primary cultures of embryonic chick retina, hormonal induction of glutamine synthetase, a glial specific enzyme, requires contact with neurons (Linser and Moscona, 1983). Contact interaction with sensory neurons is also necessary for the induction of S-100 protein by glial precursor cells in dorsal root ganglion cultures (Holton and Weston, 1983). Immunocytochemical studies show that the levels of sn-glycerol-3-phosphate dehydrogenase, an enzyme present at high levels in Bergmann glia from normal mouse cerebellum, were reduced in three neurological mutants (lurcher, nervous and purkinje cell degeneration) which show either a partial or a complete loss of Purkinje cells (Fisher, 1984).

Neurons have also been shown to affect glial cell proliferation and morphological differentiation. When sensory neurons are plated with Schwann cells, only cells in contact with fibers undergo proliferation while Schwann cells located away from the fibers do not (Wood and Bunge, 1975; Varon and Bunge, 1978). Moreover, neuronal and axolemmal membranes have been reported to stimulate Schwann cell mitosis (Hanson and Partlow, 1977; Salzer et al., 1980). The mitotic activity and shape of retinal glial cells appears to depend on the presence of neurons (Folkman and Moscona, 1978). Similarly, in cerebellar cultures, morphological differentiation of astrocytes is governed by the presence of neurons (Hatten, 1983).

Finally, neuronal degeneration leads to the hypertrophy and proliferation of glia cells located around the sites of degeneration (Duffy, 1985). A concomitant increase in the intermediate filament content is also seen in astrocytes in these cases.

We are interested in the influence of neurons on gene expression in glial cells. We have chosen to study the mammalian retina as a model system, for two important reasons. First, the retina contains only six major types of neurons (photoreceptors, bipolar cells, ganglion cells, horizontal cells, amacrine cells and interplexiform cells) and two types of macroglial cells (Rodieck, 1975). The major class of retinal glia are the radially-oriented Müller cells which comprise about 90% of all glia. The second type of glia are the astrocytes which are located in the ganglion cell layer. The second reason for choice of retina is that several retinal dystrophic animals with inherited photoreceptor degeneration are readily available (LaVail, 1981).

Glial fibrillary acidic protein (GFAP) is the major constituent of the glial-specifc intermediate filaments found in differentiated astrocytes. It has a molecular weight of 50,000 daltons and is the most widely used marker for astrocytes in normal, embryonic and neoplastic tissues. The native protein can be reassembled into 10 nm filaments in vitro, and is found in phosphorylated form. GFAP appears to be highly conserved in vertebrates, and polyclonal antisera to human GFAP show broad species cross-reactivity (Eng and DeArmond, 1983; Bignami et al., 1983; Dahl et al., 1986).

INDUCTION OF GFAP IN MÜLLER CELLS

In the retina, although the two glial types, Müller cells and astrocytes, appear to be functionally similar, they show a marked difference in GFAP-immunostaining. For example, when mouse retinas are stained with a polyclonal rabbit antiserum to GFAP, labeled cell bodies and processes are found only at the level of the ganglion cell layer. These we presume are the retinal astrocytes. The radially-oriented Müller (glial) cells or the neurons remain unstained. In retinas with photoreceptor degeneration, such as the rd/rd retina, Müller cells are found to be GFAP-immunoreactive (Figure 1). Müller cell staining was also seen in the photoreceptor-deficient retinas of BALB/c mice exposed to constant light. These results suggest that photoreceptor loss resulting either from hereditary or environmental causes leads to the appearance of GFAP-immunostaining in Müller cells (Sarthy and Fu, 1987). Functional activity of photoreceptors appears not to be the reason for the

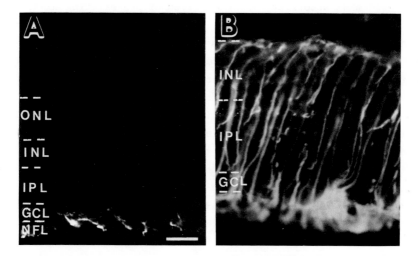

Figure 1. Immunocytochemical localization of GFAP in the mouse retina. Retinas from BALB/c mice (A) and C3H/HeJ rd/rd mice (B) were fixed in 4% paraformaldehyde and sectioned in a cryostat. 12µm sections were collected on gelatin-subbed slides, and the sections were processed for indirect immunofluorescence as described by Eisefeld et al, 1984. ONL, outer nuclear layer; INL, inner nuclear layer; IPL, inner plexiform layer; GCL, ganglion cell layer; NFL, nerve fiber layer. Scale bar: 50 µm for (A) and 10 µm for (B).

differences in staining since BALB/c mice maintained in the dark
overnight, fail to show GFAP-staining in Müller cells.

This phenomenon is not limited to mice as a similar difference
in Müller cell staining is found in other animals following retinal
degeneration or injury (Bignami and Dahl, 1979; Miller and
Oberdorfer, 1981; Dräger and Edwards, 1983; Shaw and Weber, 1983;
Eisenfeld et al., 1984; Bjorklund et al., 1985; Erickson et al.,
1987; and Eisenfeld et al., 1987).

Ultrastructural analysis of mammalian retinas shows that Müller
cells contain an abundance of intermediate (10 nm) filaments, some
of which may be the glial filaments (Rasmussen, 1972; Bussow, 1980).
Furthermore, in goldfish retina, a GFAP antiserum stains the
radially-oriented Müller cells (Bignami, 1985). Hence, one might
inquire as to why GFAP-immunostaining is not observed in Müller
cells of normal, mammalian retinas? Several explanations can be
advanced to explain this disparity. The appearance of immunostaining
in Müller cells could be due to depolymerization, proteolysis or
chemical modification of existing glial filaments, or decreased GFAP
degradation. Alternatively, it could be due to new GFAP synthesis
resulting from transcriptional activation of the GFAP gene,
increased mRNA stability, or translation of pre-existing mRNA.

We have used immunoblotting, northern blotting and in situ hybridization techniques to address this question. Immunoblotting was
carried with protein extracts from rd/rd, and light damaged (CLD)
retinas to determine whether there was a change in GFAP levels in
the photoreceptor-deficient retinas. These studies showed that there
was at least a 5-fold increase in GFAP levels in the dystrophic
retinas. Northern blots further showed that there was a 10 to 25-
fold elevation in the GFAP mRNA content in photoreceptor-deficient
retinas. Finally, in situ hybridizations with sections of retina and
isolated Müller cells established that there was at least a 10-fold
increase in GFAP mRNA levels in Müller cells from CLD-retinas
(Sarthy and Fu, 1987).

GFAP LEVELS IN ASTROCYTES

As mentioned earlier, neuronal degeneration leads to 'reactive
gliosis' in the neighboring astrocytes (Duffy, 1983).
Immunocytochemical studies with GFAP-antisera show that there is
increased immunostaining in the 'reactive' astrocytes. Hence, one

might ask what happens to GFAP levels in astrocytes from photoreceptor-deficient retinas? This question is particularly interesting since astrocytes are located far away from photoreceptors and have no direct contact with them.

Unfortunately, because of the intense staining present in the end feet region of Müller cells, it is impossible to determine if there is an increase in GFAP-immunostaining in astrocytes. This problem can be circumvented by looking at GFAP mRNA levels in astrocytes of normal and CLD-retinas by in situ hybridization. Since Müller cell bodies are located in the inner nuclear layer while the astrocyte somata are present in the nerve fiber layer, any contribution from Müller cells should be minimal. A comparison of the silver grain density on labeled astrocytes suggested that there was a 2 to 3-fold increase in GFAP mRNA levels in astrocytes from CLD-retinas (Sarthy and Fu, 1987).

BASIS FOR INCREASED GFAP-STAINING

Although astrocytic gliosis is common to many kinds of CNS trauma and disease, the mechanism responsible for increased GFAP-immunostaining remains uncertain. In experimental autoimmune encephalomyelitis (EAE) in Lewis rats, Smith et al. (1985) found enhanced GFAP-staining in astrocytes, 10-12 days after inoculation. ^{3}H-amino acid labeling experiments showed that there was a substantial increase in GFAP synthesis as the disease became acute, and GFAP levels decreased as the animals recovered. In contrast, Goldmuntz et al. (1987) reported that glial reaction in EAE was not accompanied by significant changes in GFAP synthesis, although the synthesis of vimentin was elevated. Increase in GFAP-staining has been found as early as half hour following cryogenic lesion (Amaducci et al., 1981). Recently, Lipsky and Freese (1987) reported that in a human astrocytoma cell line, GFAP synthesis could be induced in the presence of the protein synthesis inhibitor, cyclohexamide. They have proposed that turnover of an unidentified protein may be involved in GFAP induction in this case. In C6 glioma cells induced with dibutyryl cAMP, maximal GFAP levels were observed when the net protein synthesis was drastically depressed (Backhovens et al., 1987).

In our studies, we found a 5-fold increase in GFAP levels in retinas with photoreceptor degeneration. We presume that this increase in GFAP content results in the appearance of immunostaining in Müller cells. Furthermore, the increased GFAP level is likely to be due to new synthesis rather than being due to decreased degradation as intermediate filaments are extremely stable (Lazarides, 1982). Our findings are in agreement with results from detached cat retina (Erickson et al., 1987) and dystrophic rat retina (Eisenfeld et al., 1984) in which cases increases in GFAP content were also accompanied by the appearance of GFAP-immunostaining in Müller cells. In addition, our northern blotting data and in situ hybridization results prove that there is increased GFAP gene transcription in retinal glia.

CELL INTERACTIONS AND GFAP INDUCTION

Our findings as well as the results obtained by other investigators suggest that photoreceptor loss leads to GFAP expression in Müller cells. The cellular mechanism behind this phenomenon is, however, unclear. One possibility is that some aspect of photoreceptor-Müller cell interaction keeps the GFAP gene repressed and that loss of photoreceptors leads to GFAP induction. In support of this possibility, when embryonic retinal cells are dissociated and plated in vitro, the glial cells in these cultures show GFAP-staining (Pech and Sarthy, 1983). It could also be argued that degenerating photoreceptor membranes induce GFAP synthesis. Alternatively, the lymphokine and monokine factors produced by invading macrophages may be involved in GFAP gene activation (Figure 2). For instance, ConA-stimulated lymphocytes have been reported to release 'factors' that stimulate GFAP synthesis in astrocytes (Fontana et al., 1981a,b).

Our observation that photoreceptor degeneration leads to activation of retinal astrocytes suggests that contact with degenerating photoreceptors may not be necessary and that diffusible factors may be involved in GFAP induction in astrocytes. It could, however, be argued that activation of astrocytes occurs through 'reactive' Müller cells with which they are in contact at the inner limiting membrane. It might be pointed out here that GFAP expression in Müller cells is also observed in other cases such as retinal detachment, retinal injury, uveoretinitis or optic nerve injury (Bignami and Dahl, 1979; Miller and Oberdorfer, 1981; Shaw and Weber, 1983; Eisenfeld et al, 1984; Bjorklund et al., 1985; Erickson

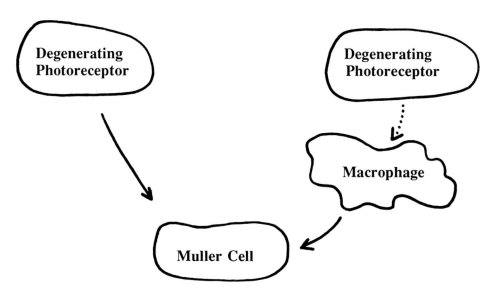

Figure 2. Two possible mechanisms for GFAP induction in Müller cells. In the first mechanism, membranes or other cellular components derived from degenerating photoreceptors, act directly on Müller cells. In the second mechanism, macrophages that invade retina as a result of photoreceptor degeneration, release factors that stimulate GFAP synthesis in Müller cells.

et al., 1987; and Eisenfeld et al., 1987). It is possible that GFAP induction in all these cases occurs through a common cellular mechanism.

Finally, one might inquire as to the functional significance behind GFAP induction in retinal glia. Since we do not know the cellular function of intermediate filaments, it is difficult to understand the requirement for GFAP in Müller cells. It is quite possible that GFAP expression is adventitious and is the result of induction of an unidentified gene in Müller cells. Since the phenomenon is of wide occurrence, being observed in all mammalian retinas examined so far, we feel that GFAP induction in Müller cells is likely to have some functional significance.

In summary, it appears that retinal degeneration resulting from a wide variety of causes such as mutations, light damage, retinal detachment and autoimmune reaction, leads to GFAP induction in Müller cells in the mammalian retina. The molecular and cellular mechanisms that underlie this phenomenon remain to be elucidated.

ACKNOWLEDGEMENTS

During preparation of this review, the author was supported by NIH grants, EY-02324, EY-03664, EY-01730, and unrestricted funds from the Research to Prevent Blindness Foundation. I wish to thank Mr. Daniel Possin for help in preparing the manuscript.

REFERENCES

Amaducci, L., K.I.Forno and L.F. Eng. 1981. Glial fibrillary acidic protein in cryogenic lesions of the rat brain. Neurosci. Lett. 21: 27-32.

Backhovens, H., J. Gheuens and H. Slegers. 1987. Expression of Glial fibrillary acidic protein in rat C6 glioma relates to vimentin and is independent of cell-cell contact. J. Neurochem. 49:348-354.

Bignami, A. 1984. Glial fibrillary acidic (GFA) protein in Müller glia. Immunofluorescence study of the goldfish retina. Brain Res. 300:174-178.

Bignami, A. and D. Dahl. 1979. The radial glia of Müller in the rat retina and their response to injury. An immunofluorescence study with antibodies to the glial fibrillary acidic (GFA) protein. Exp. Eye Res. 28:63-69.

Bignami, A., L.F. Eng, D. Dahl and C.T. Uyeda. 1972. Localization of the glial fibrillary acidic protein in astrocytes by immunofluorescence. Brain Res. 43:429-435.

Bignami, A., D. Dahl, and D.G. Rueger. 1980. Glial fibrillary acidic (GFA) protein in normal neural cells and in pathological conditions. Adv. Cell. Neurobiol. 2:85-310.

Bjorklund, H., A. Bignami and D. Dahl. 1985. Immunohistochemical demonstration of glial fibrillary acidic protein in normal rat Müller glia and retinal astrocytes. Neurosci. Lett. 54:363-368.

Bray, G.M., M. Rasminsky and A.J. Aguayo. 1981. Interactions between axons and their sheath cells. Ann. Rev. Neurosci. 4:127-162.

Bunge, R.P., M.B. Bunge and C.F. Eldridge. 1986. Linkage between axonal ensheathment and basal lamina production by Schwann cells. Ann. Rev. Neurosci. 9:305-328.

Bussow, H.1980. The astrocytes in the retina and optic nerve head of mammals: A special glia for the ganglion cell axons. Cell Tissue Res. 206:367-378.

Dahl, D., H. Borklund and A. Bignami. 1986. Immunological markers in astrocytes. In Astrocytes. Cell Biology and Pathology of Astrocytes. Vol. 3. Eds. Federoff, S. and A. Vernadakis. Academic Press, Inc., Orlando, FL. pp. 1-25.

Dräger, U.C. and D.L. Edwards. 1983. Antibodies to intermediate filaments reveal abnormalities in retinas of mice with photoreceptor degeneration. Invest. Ophthalmol. Vis. Sci. 24:115.

Duffy, P.E. 1983. Astrocytes: normal, reactive and neoplastic. Raven Press, New York.

Eisenfeld, A.J., A. Bunt-Milam and P.V. Sarthy. 1984. Müller cell expression of glial fibrillary acidic protein after genetic and experimental photoreceptor degeneration in the rat retina. Invest. Ophthal. Vis. Sci. 25:1321-1328.

Eisenfeld, A.J., A.H. Bunt-Milam and J.C. Saari. 1987. Uveoretinitis in rabbits following immunization with interphotoreceptor retinoid-binding protein. Exp. Eye Res. 44:425-438.

Eng, L.F., and S.J. DeArmond. 1982. Immunocytochemical studies of astrocytes in normal development and diseases. Adv. Cell. Neurobiol. 3:145-171.

Erickson, P.A., S.K. Fisher, C.J. Guerin, D.H. Anderson and D.D. Kaska. 1987. Glial fibrillary acidic protein increases in Müller cells after retinal detachment. Exp. Eye Res. 44:37-48.

Fisher, M. 1984. Neuronal influence on glial enzyme expression: Evidence from mutant mouse cerebella. Proc. Natl. Acad. Sci. 81:4414-4418.

Folkman, J. and A.A. Moscona. 1978. Role of cell shape in growth control. Nature (London) 273:345-349.

Fontana, A., R. Dibs, R. Merchant, S. Balsiger and P.J. Grob. 1981a. Glial cell-stimulating factor (GSF): a new lymphokine. Part 1. Cellular sources and partial purification of murine GSF, role of cytoskeleton and protein synthesis in its production. J. Neuroimmunol. 2:55-71.

Fontana, A., U. Otz, A.L. DeWeck, and P.J. Grob.1981b. Glia cell stimulating factor (GSF): a new lymphokine. Part 2. Cellular sources and partial purification of human GSF. J. Neuroimmunol. 2:73-81.

Goldmuntz, E.A., C.F. Brosnan, F.-C. Chiu, and W.T. Norton. 1986. Astrocytic reactivity and intermediate filament metabolism in experimental autoimmune encephalomyelitis: The effect of suppression with prazosin. Brain Res. 397:16-26.

Hanson, G.R. and L.M. Partlow. 1978. Stimulation of non-neuronal cell proliferation in vitro by mitogenic factors present in highly purified sympathetic neurons. Brain Res. 159:195-210.

Hatten, M.E. 1985. Neuronal regulation of astroglial morphology and proliferation in vitro. J. Cell Biol. 100:384-396.

Hiscott, P.S., I. Grierson, C.J. Trombetta, A.H. Rahi, J. Marshall and D. McLeod. 1984. Retinal and epi-retinal glia-an immunohisto-chemical study. Br. J. Ophthalmol. 68:698-707.

Holton, B. and J.A. Weston. 1982. Analysis of glial cell differentiation in peripheral nervous tissue. II. Neurons promote S-100 synthesis by purified glial precursor cell populations. Dev. Biol. 89:72-81.

LaVail, M.M. 1981. Analysis of neurological mutants with inherited retinal degeneration. Invest. Ophthal. Vis. Sci. 21:638-668.

Lazarides, E. 1982. Intermediate filaments: A chemically heterogeneous, developmentally regulated class of proteins. Ann. Rev. Biochem. 51:219-250.

Lees, M.B. and S.W. Brostoff. 1984. In Myelin, 2nd edition. Ed. Morell, P. Plenum Press, Inc., pp. 197-224.

Linser, P. and A.A. Moscona, A.A. 1983. Hormonal induction of glutamine synthetase in cultures of embryonic retinal cells: requirement for neuron-glia contact interactions. Dev. Biol. 96:529-539.

Lipsky, R.H. and E. Freese. 1987. Induction of gfa gene expression for glial fibrillary acidic protein (GFAP) in a human astrocytoma-derived cell line. Sec. World Cong. Neurosci. 2:117.

Miller, N.M. and M. Oberdorfer. 1981. Neuronal and neuroglial responses following retinal lesions in the neonatal rats. J. Comp. Neurol. 202:493-504.

Molnar, M.L., K. Stefansson, L.S. Marton, R.C. Tripathi and G.K. Molnar. 1984. Exp. Eye Res. 38:27-34.

Pech, I.V. and P.V. Sarthy. 1984. Development of GABAergic neurons in primary cultures of rat retina: comparison of neuronal and mixed cultures. Soc. Neuosci. Abst. 9:942.

Purves, D. and J.W. Lichtman. 1985. Principles of Neural Development. Sinauer Associates Inc., Sunderland, MA.

Rasmussen, K.-E. 1972. A morphometric study of the Müller cell cytoplasm in the rat retina. J. Ultrastruct. Res. 39:413-429.

Rodieck, R. W. 1973. The Vertebrate Retina. Principles of structure and function. W.H. Freeman and Co., San Franscisco, CA.

Salzer, J.L., A.K. Williams, L. Glaser and R.P. Bunge. 1980. Studies of Schwann cell proliferation. II. Characterization of the stimulation and specificity of the response to a neurite membrane fraction. J. Cell Biol. 84:753-766.

Sarthy, P.V. and M. Fu. 1987. Expression of the glial fibrillary acidic protein (GFAP) gene in the mouse retina. Soc. Neurosci. Abst. 13:378.

Shaw, G. and K. Weber. 1983. The structure and development of the rat retina: an immunofluorescence microscopical study using antibodies specific for intermediate filaments proteins. Eur. J. Cell Biol. 30:219-232.

Smith, M.E., F.P. Somera and L.F. Eng. 1983. Immunocytochemical staining for glial fibrillary acidic protein and the metabolism of cytoskeletal proteins in experimental allergic encephalomyelitis. Brain Res. 264:241-253.

Varon, S. S., and Bunge, R.P. 1978. Trophic mechanisms in the peripheral nervous system. Ann. Rev. Neurosci. 1, 327-361.

Varon, S. and G.G. Somjen. 1979. Neuron-glia interactions. Neurosci. Res. Prog. Bull. 17:1-239.

Wood, P.M. and R.P. Bunge. 1975. Evidence that sensory axons are mitogenic for Schwann cells. Nature (London) 256:662-664.

Müller Cells Guide Migrating Neuroblasts in the Developing Teleost Retina
Pamela A. Raymond

The list of situations in which glia are thought to help organize neurons during development is growing. Glia have been implicated in such diverse processes as neuronal migration, the formation of cytoarchitectonic patterns, and the guidance of axonal outgrowth both within and outside the central nervous system. Glial cells apparently assume this role very early in development, probably from the time of their origin. The earliest formed are the radial glia, bipolar cells with long, thin processes that span the entire width of the neural tube. Classic studies by Ramon y Cajal (1911) using the Golgi technique, later confirmed by Rakic, Sidman and others (Rakic, 1971; Rakic, 1972; Sidman and Rakic, 1973), suggested that these early radial glia provide a physical pathway along which young neurons migrate toward the pial surface from their site of origin near the ventricle. Each migrating neuron is tightly apposed to the surface of a radial glial fiber and has a short leading process that spirals around the fiber (Rakic, 1971). Hatten and colleagues (Hatten, et al., 1984) have dissociated and cultured cells from neonatal mouse cerebellum, and in their cultures cerebellar granule neurons can be seen attaching to and migrating along the extended processes of Bergmann (radial) glia.

Several investigators have proposed that glia not only guide migrating neuronal cells, they also are involved in directing or channeling the outgrowth of their axons. For example, a specialized glial structure in the embryonic forebrain supports growing callosal fibers as they cross from one side to the other (Silver, et al., 1982). In other situations it is thought that glia block or impede growing axons (Smith, et al., 1986; Silver, et al., 1987). Recent work by Cooper and Steindler (1986) suggests that glia may also be involved in the partitioning of axon terminals and their postsynaptic target neurons into functional subregions within a given brain region, such as the barrel fields within the somatosensory cortex (S1) of the mouse. Still another suggestion is that glia regulate synaptic density by ensheathing the dendritic surface; when the glial coverings are removed, the neurons become hyperinnervated (Meshul, et al., 1987).

What these diverse examples all have in common is the idea that glia not only provide the structural framework of the nervous system, they also help to create it. Many of the recent discoveries in this field have used the cerebellum as a model system. The retina provides another opportunity to study the role of glia in organizing neuronal architecture during development. Like the cerebellum, the retina has relatively few neuronal types, and these are partitioned into discrete laminae that run parallel to the surface. Also like the cerebellum, the retina has its own specialized version of radial glia, the Müller cell (Ramon y Cajal, 1892), whose long axis is perpendicular to the plane of the cellular strata. Different types of retinal neurons are located in different strata, and Sheffield and Li (1987) have proposed that Müller cells might themselves be stratified in ways that define the different layers, raising the possibility that they may direct migrating neurons to their proper levels during retinal histogenesis. Recent evidence from my laboratory has shown that Müller cells do provide a substrate for migrating neuroblasts in the postembryonic teleost retina (Raymond and Rivlin, 1987). To summarize briefly, these migrating cells are mitotically active progenitors of rod photoreceptors, and they are initially sequestered in the inner nuclear layer when the retinal laminae first form in the embryo (Johns, 1982; Raymond and Rivlin, 1987). Later, in larval fish, these cells move across the outer plexiform layer into the outer nuclear layer (Raymond and Rivlin, 1987), where they continue to proliferate and produce new rods even in adult fish (Sandy and Blaxter, 1980; Johns and Fernald, 1981). While in the inner nuclear layer and during their journey across the outer plexiform layer, the migrating rod precursors are closely associated with the radial fibers of Müller glia (Raymond and Rivlin, 1987). I will next review the evidence that led us to propose this idea.

RETINAL HISTOGENESIS

Let me start with a brief review of histogenesis in the vertebrate retina (Mann, 1928; Raymond, 1985), as diagrammed in figure 1. The retinal primordium starts as a primitive neuroepithelium. All mitotic divisions are at the external limiting membrane (the former ventricular surface) and processes of the neuroepithelial cells span the epithelium during interphase of the cell cycle (figure 1A). Histogenesis begins with the orderly withdrawl of neurons from the mitotic cycle. Ganglion cells are first (figure 1B) and other neurons soon follow, as the neuroepithelium is partitioned into three layers. These layers are not identical in cell content to those of the mature retina, however. For example, the early outer nuclear layer is composed of single row of cone nuclei (figure 1C). Rods are the last retinal neurons to be born (figure 1D). During larval stages in fish, mitosis continues in neuroblastic cells in the outer part of the inner nuclear layer, and these cells later migrate into the outer nuclear layer where they contribute to the ongoing generation of new rods, as described above.

AUTORADIOGRAPHIC EVIDENCE FOR INTERLAMINAR MIGRATION DURING RETINAL DEVELOPMENT

Figure 2A shows proliferating cells in the inner nuclear layer of the larval goldfish retina. A pulse of ^{3}H-thymidine was given on the day after hatching, and the retina was prepared for autoradiography 4 days later. Note the clusters of lightly labeled, elongated cells in the inner nuclear layer. The dilution of label indicates that these cells had continued to divide during the interval following injection. The cluster on the right spans the outer plexiform layer and is contiguous with a labeled cell in the outer nuclear layer. Another group of fish was injected with ^{3}H-thymidine at a similar age, that is, during first few days after hatching, but their retinas were prepared for autoradiography within 24 hours. In those preparations all labeled cells were in the inner nuclear layer, and none were in the outer nuclear layer (Johns, 1982). At these short survival times only dividing cells are labeled, so we can conclude that there were no dividing cells in the outer nuclear layer at hatching, but labeled nuclei appeared there later. At least some of the labeled nuclei that later appeared in the outer nuclear layer differentiated into rods. This was shown by electron microscopic autoradiography of retinas from larval fish injected with ^{3}H-thymidine and allowed to survive for a week or more (figure 2B). At these longer survival times, the cells that remain labeled are those that incorporated the label and shortly thereafter withdrew from the mitotic cycle and began to differentiate. Cells that continued dividing were no longer labeled under these circumstances, because the ^{3}H-thymidine was diluted with each successive round of DNA synthesis. The labeled cells in these preparations were almost without exception differentiating rods (Raymond and Rivlin, 1987). These observations suggest that dividing cells in the inner nuclear layer, or their postmitotic progeny, must have migrated into the outer nuclear layer, where they differentiated into rods.

In contrast to the situation in the early larval fish, when ^{3}H-thymidine was given to older larvae or juveniles, most labeled dividing cells were in the outer nuclear layer (figure 2C). Only a few remained in the inner nuclear layer (Johns, 1982; Raymond and Rivlin, 1987). The relative lack of labeled dividing cells in the inner layers in older animals and their simultaneous appearance in the outer nuclear layer is consistent with the proposal that undifferentiated cells leave the former and move to the later and once there, continue to divide. Some of the progeny of these cells differentiate into rods, but others must continue to proliferate,

Figure 1. Retinal histogenesis in vertebrates. A, the primitive neuroepithelium is composed of dividing, undifferentiated neuroepithelial cells. Mitotic figures (v) are at the ventricular surface (external limiting membrane in the retina). B, the first neurons to withdraw from the mitotic cycle are ganglion cells, and they move away from the ventricular surface and congregate in the presumptive ganglion cell layer (G). C, with further development, the neuroepithelium is partitioned into the 3 definitive cellular strata of the retina, ganglion cell layer (G), the inner nuclear layer (IN) and the outer nuclear layer (ON). At this early stage, the ON consists of a single row of immature cones, in both teleost fish (Blaxter and Jones, 1967; Blaxter and Staines, 1970; Johns, 1982) and mammals (Mann,1928). The few small dark nuclei at the base of the cones are presumptive rods. D, in the mature retina of most teleost fish there is a single row of cone nuclei (C) and multiple rows of rod nuclei (R) in the outer nuclear layer. A similar situation holds in most mammalian retina, and in the non-foveal regions of the primate retina. Reprinted with permission from Trends in Neurosci. 8:12 (1985).

15

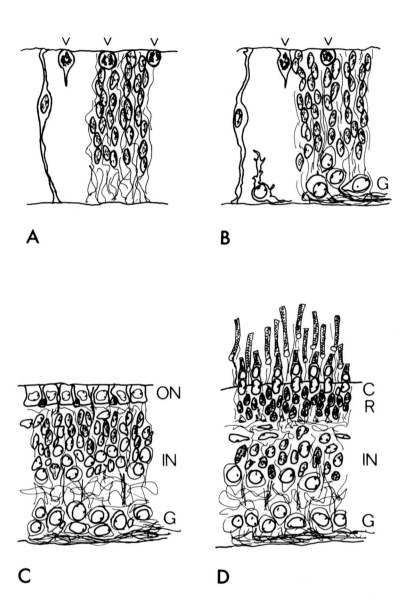

A   B

C   D

since rod precursors are found in the outer nuclear layer of adult fish (Johns and Fernald, 1981). Quantification of labeled retinal cells in fish injected at different ages and allowed to survive for varying periods after the injection provided evidence in support of this interpretation (figure 3). In the youngest fish, there were no dividing cells (represented as open circles in figure 3) in the outer nuclear layer, that is, no labeled cells were found there at survival times less than 24 hours. In increasingly older fish, a greater fraction of dividing cells were in the outer nuclear layer (30% in late larvae and 80-90% in juvenile and adults). At all stages, with increased survival time (filled circles), labeled cells disappeared from the inner and accumulated in the outer nuclear layer. For 2 fish injected at 1-2 months of age, one eye was removed at 24 hours and the other eye either 30 or 110 days later; values from the two retinas are connected by vertical lines (figure 3). In both cases, there was a higher fraction of labeled cells in the outer nuclear layer in the second eye. Since we could not look at the same retina twice, this is the closest we could come to demonstrating an actual shift in the apportionment of labeled cells between retinal layers. We concluded that mitotic progenitors of rods migrate vertically across retinal layers during postembryonic development in goldfish, and that the dividing cells in the inner nuclear layer of the early larval fish apparently do not make a permanent contribution of new cells to that layer. The only other way to account for the disappearance of labeled cells from the inner

Figure 2. Autoradiographs of ^{3}H-thymidine-labeled cells in goldfish retina. A, this light micrograph is from a larval fish injected 4 days previously with ^{3}H-thymidine. Note the clusters of spindle-shaped labeled cells in the inner nuclear layer (arrows). The one on the right is continuous across the outer plexiform layer (OP) into the outer nuclear layer. The relatively light label (compare to 2C) indicates that these cells had divided several times during the interval after the ^{3}H-thymidine was administered. Other abbreviations as in figure 1. Calibration bar, 20µm. B, this electron micrograph shows 2 developing rods (arrowheads) in central retina of a larval goldfish. ^{3}H-thymidine was injected 9 days previously; the label over the nuclei of these young rods indicates that they underwent their terminal mitotic division soon after the label was administered. Calibration bar, 5 µm. C, this light micrograph shows rod precursors in the outer nuclear layer of a juvenile goldfish. These cells were labeled by an injection of ^{3}H-thymidine 4 hours previously, indicating that they were mitotically active. Calibration bar, 20 µm. A and C reprinted with permission from J. Neurosci. 2:178 (1982); B reprinted with permission from Develop. Biol. 122:120 (1987).

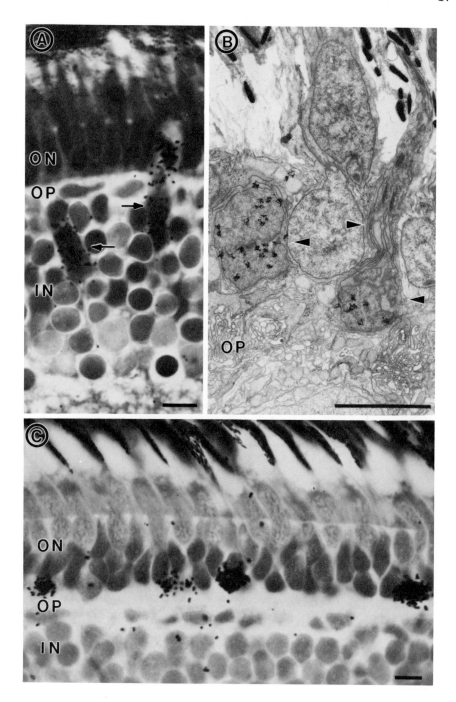

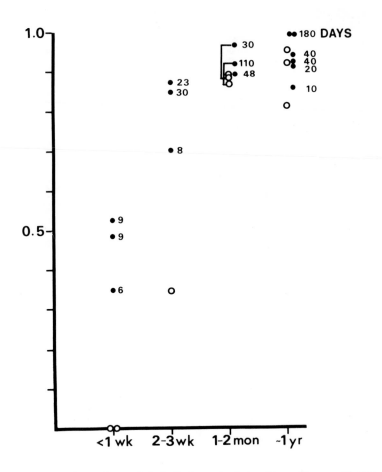

Figure 3. The fraction of labeled nuclei in the outer nuclear layer of larval and juvenile goldfish. Labeled nuclei in the inner and the outer nuclear layers were counted across the extent of the retina. In larval fish, all sections that passed through the lens in a set of serial sections through the head were scanned for labeled nuclei. In the larger eyes of juvenile fish (1 to 2 months old), nuclei were counted at 15 μm intervals in the set of sections that passed through the lens. In eyes from 1-year-old fish, seven parameridional sections within an 80 μm interval were counted. The fraction of labeled nuclei in the outer nuclear layer (ONL), compared to the total labeled nuclei (outer plus inner nuclear layers)(ordinate) is plotted as a function of the age of the fish (abcissa) at injection. Each point represents one retina. The open circles are from eyes which were removed 1 day or less after injection. The solid circles are from eyes removed at longer intervals; the length of the survival period in days is given by the number to the right of each point. In two of the fish injected at 1 to 2 months of age, the right eye was removed at one day, and the left eye was removed several days later. Counts from the two eyes from the same fish are connected by vertical line segments. Reprinted with permission from J. Neurosci. 2:178 (1982).

nuclear layer is to invoke cell death. However, in examining a series of autoradiographic preparations with electron microscopy we found thymidine-labeled dying cells in the outer nuclear layer, but not in the inner (Raymond and Rivlin, 1987).

MORPHOLOGICAL EVIDENCE FOR MIGRATION OF NEURONAL CELLS ALONG RADIAL FIBERS OF MULLER GLIA

The ^{3}H-thymidine data suggested that cells were migrating across retinal layers, but is there any morphological evidence to suggest that this happens? When we examined with the electron microsope retinas from larval goldfish injected with ^{3}H-thymidine, we found that labeled cells in the inner nuclear layer were invariably closely associated with Müller cells and their processes (Raymond and Rivlin, 1987). Müller cells were recognized by their polygonal, deeply clefted nuclei and the high content of glycogen in their cytoplasm. As we had seen in autoradiographs from the light microscopic study, the labeled cells were typically arranged in clusters. The micrograph in figure 4A shows a cluster of dark cells in the inner nuclear layer. These are immature, dividing cells as revealed by their incorporation of label in an adjacent section that was processed for autoradiography (figure 4B). The labeled cells are electron dense and have a spindle shaped nucleus. Note the adjoining (unlabeled) Müller cell nucleus. To examine the morphology of these cells and their association with Müller glia, we used serial reconstruction techniques. We prepared three serial sets of 150 to 250 sections each from 2 larval retinas labeled with ^{3}H-thymidine. Both retinas were prepared less than 24 hours after injection so that the label was only over dividing cells. About every 20th section in the series was processed for autoradiography. Figure 5 shows the reconstruction of the labeled cluster shown in figure 4. These labeled cells were tightly apposed to the Müller processes and twisted around them in a way that is very reminiscent of the migrating neuroblasts in neocortex and cerebellum (Rakic, 1971, 1972). Note that the cluster of labeled cells in figures 4 & 5 spans the outer plexiform and reaches into the outer nuclear layer. This was true of some, but not all the clusters we observed, and is consistent with our proposal that they were migrating, since cells with this shape were never found in the outer plexiform layer in the mature goldfish retina (Johns, 1982).

Figure 6 summarizes the results from our serial reconstructions. There are several important points. The streamlined shape of dividing cells in the inner nuclear layer is consistent with a migratory behavior, whereas the globular shape of those in the outer nuclear layer may reflect a more sedentary existence. We have labeled them P1 and P2, respectively. The "P" stands for "precursor" or "progenitor" and the numbers signify what we believe to be a temporal relationship: 2 is derived from 1. The dividing cells did not retain a connection with the ventricular surface (ELM), and mitotic figures were seen in both inner and outer nuclear layers. A mitotic figure in the outer nuclear layer is indicated by the cell with an asterick. Thus, these cells differ from true neuroepithelial cells of the embryonic retina, which do not relinquish their hold on the ventricular surface until they withdraw from the mitotic cycle. In the terminology of the Boulder Committee (1970), these cells are equivalent to subventricular cells which are found in some regions of the developing brain and which give rise to neurons born late in development. This pattern fits that described for the retina, in that these dividing cells are found in postembryonic fish and give rise to rods, the last neurons to be generated (Blaxter and Jones, 1967; Blaxter and Staines, 1970; Johns, 1982; Raymond, 1985). Finally, the P1 precursors in the inner nuclear layer are closely associated with Müller glia, and we suggest that they may use the glial fibers to guide them across the outer plexiform layer into the outer nuclear layer. Dividing cells in the outer nuclear layer persist even in adult fish. The progeny of these cells differentiate into rods and only rods (Johns and Fernald, 1981; Raymond and Rivlin, 1987), and we have therefore

Figure 4. Electron micrographs of a cluster of labeled dividing cells in the inner nuclear layer of a larval goldfish. The fish was injected with ^{3}H-thymidine 12 hours before preparation of the tissue. A, a group of 4 elongated, electron dense cells (P) associated with a Müller cell (M). The cluster of cells spans the outer plexiform layer (OP) from inner nuclear (IN) to outer nuclear (ON) layers. The visual processes of young cone photoreceptors (PR) can be seen near the top. B, a nearby section in the serial set which included the section illustrated in A. The section in B was processed for autoradiography. The 4 cells in the cluster have silver grains over their nuclei, indicating that they were mitotically active. The Müller nucleus is not labeled. Calibration bars, A and B, 5 μm. Reprinted with permission from Develop. Biol. 122:120 (1987).

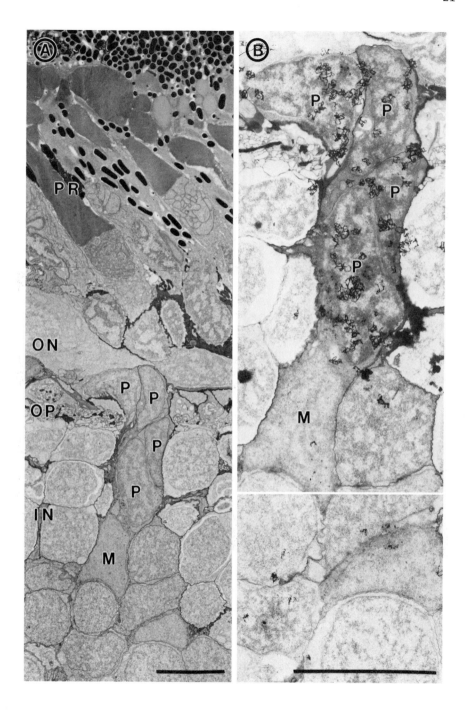

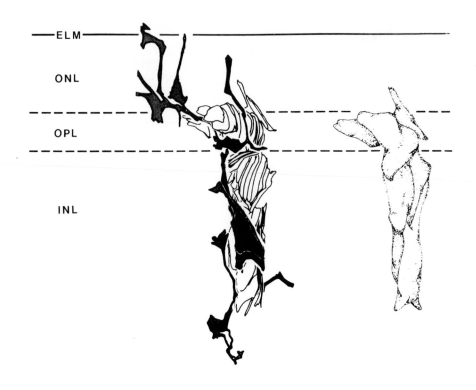

ELM

ONL

OPL

INL

Figure 5. Computer generated reconstruction of the cluster of labeled cells shown in figure 4. The Müller cell and some (not all) of its processes are indicated in dark grey; the labeled cells were reconstructed from outlines of every 5th section. On the right side is an artist's rendition of the labeled cluster, without the Müller processes. The cluster was composed of 6 cells, only 4 of which could be seen in the sections illustrated in figure 4. Abbreviations: external limiting membrane (ELM), outer nuclear layer (ONL), outer plexiform layer (OPL), inner nuclear layer (INL). Modified from Develop. Biol. 122:120 (1987).

named them rod precursors. An immature rod is shown at the far right in figure 6.

What about the dividing cells that persist in the inner nuclear layer of postembryonic fish, albeit in ever shrinking proportions relative to rod precursors in the outer nuclear layer? These remain an enigma. The longitudinal thymidine studies made it clear that they make no permanent or lasting contribution to the layer in which they reside, which to the contrary is continually losing cells (Blaxter and Jones, 1967; Johns, 1982). They may function simply as a reserve of undifferentiated cells, waiting to seed the outer

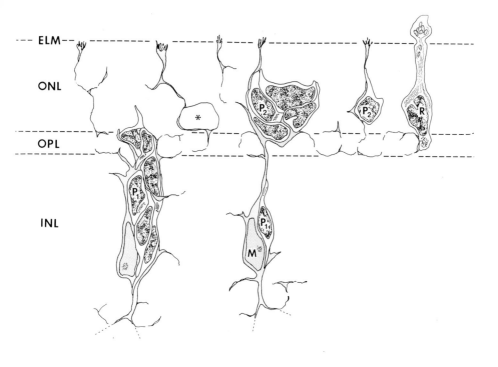

Figure 6. Semischematic diagram illustrating the results of the serial reconstructions. The drawings are arranged from left to right in a proposed sequence leading to the ontogenesis of rods in the goldfish retina (see text for details). Abbreviations: mitotic figure (*), immature rod (R), other abbreviations as in figures 4 and 5. Reprinted with permission from Develop. Biol. 122:120 (1987).

nuclear layer with rod precursors. The clear association of these cells with Müller glia raises the obvious question whether some of these cells might be dividing Müller cells. It is thought that Müller cells, like many glia, retain the capacity to proliferate especially under traumatic conditions (Nork, et al., 1987; Wilson, et al., 1987), and they are known to proliferate in vitro (Kaplowitz and Moscona, 1976). However, we do not believe that Müller cells in the postembryonic goldfish retina proliferate under normal conditions. In our electron microscopic autoradiographic study we examined over 200 individual thymidine-labeled cells in retinas from 1- to 2-month-old fish, and we saw not a single labeled Müller cell (Raymond and Rivlin, 1987). However, we do not know when Müller cells are born in the fish retina. One intriguing possibility is that the P1 precursors in a given cluster and the Müller cells with which they are so intimately associated are clonally related, that is, they arose as the division products of a single grandmother cell. This possiblity is suggested by the recent work of Cepko and her colleagues (Turner and Cepko, 1987; Price, et al., 1987), who have developed an exciting new method for lineage analysis. They infected retinas of neonatal rats with a replication incompetent retrovirus which had been genetically engineered with recombinant technology so that it carried the gene for a bacterial enzyme, B-galactosidase. These altered viruses infected proliferating retinal cells in the host, where they were incorporated into the genome and thus provided a heritable marker whose presence was later revealed with histochemical techniques. At a low dilution of virus, the infected cells were widespread enough that unique labeled clones could be identified. The clones they found were comprised of several different cell types, but most contained rods, which are the last cells generated in the rat retina as in fish. Some contained one or more rods and a Müller cell. The clear implication of these results is that there are not separate progenitor cells for different types of retinal neurons, and in particular, neurons and Müller cells can share a common progenitor. If a similar situation holds in fish retinas, then it might imply that the clusters of dividing cells in the inner nuclear layer and the Müller cells associated with them are clonally related.

CELL LINEAGE RELATIONSHIPS IN THE TELEOST RETINA: A PROPOSAL

Our current working hypothesis for the lineage relations of retinal cells in the fish retina is shown in figure 7. Primitive neuroepithelial cells give rise sequentially to various types of neurons. The exact order is not certain, and may not be absolute. One can generalize by saying that ganglion cells are born first and rods last. Müller glia may be generated by the terminal division of a neuroepithelial cell, giving rise to a postmitotic Müller cell and a P1 precursor. The latter cells spawn the line that produces rod precursors and ultimately rods. Such a clonal relation between Müllers and P1 precursors is appealing because it would provide a convincing explanation for the close association of these cell types in the larval retina. However, this lineage pattern would put the

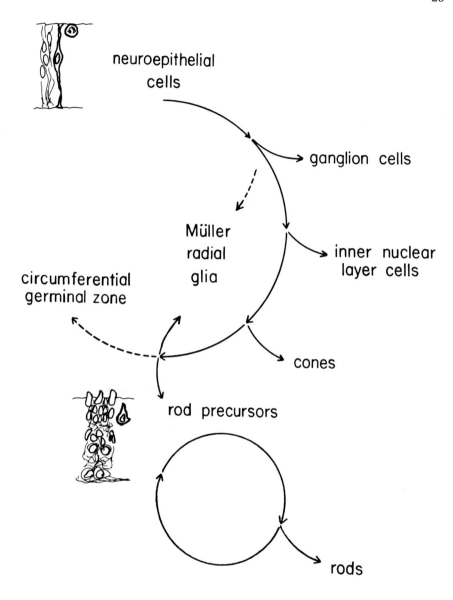

neuroepithelial
cells

ganglion cells

Müller
radial
glia

inner nuclear
layer cells

circumferential
germinal zone

cones

rod precursors

rods

Figure 7.   Diagram to show a proposed ontogenetic history of teleost retinal neurons.   Modified from Trends in Neurosci. 8:12 (1985).

birthdate of Müller cells rather late in development, and because there is some evidence from other species that Müller cells are born early in retinal development (Mann, 1928; Uga and Smelser, 1973; Kuwabara and Weidman, 1974), we have included a second possible origin at an earlier stage. I must emphasize that these ideas about Müller cell origins are only speculative. We are presently investigating the issue of Müller cell birthdates in the embryonic goldfish retina.

There is one final point to be made about figure 7. The termination of the neuroepithelial cell cycle in the circumferential germinal zone refers to the fact that in the teleost retina there is a growth zone at the ciliary margin where production of annuli of new retina by neuroepithelial cells continues into adulthood (Raymond, 1985). Although I have not discussed this feature of postembryonic growth, let me just add that in this peripheral growth zone, the stages described here for histogenesis of the larval retina are exactly recapitulated.

POSSIBLE RELEVANCE OF THESE RESULTS TO RETINAL DEVELOPMENT IN OTHER SPECIES

In summary, these studies have provided evidence for Müller cell involvement in the migration of neuronal germinal cells across retinal laminae in the postembryonic teleost retina. Fish are unique in that cell proliferation continues in the retina throughout postembryonic life, both at the circumferential germinal zone and in the scattered population of rod precursor cells (Raymond, 1985). However, late genesis of rod photoreceptors is also a property of the developing mammalian retina. Carter-Dawson and LaVail, (1979) showed that most rods in the mouse retina are generated postnatally, after production of cones ceases. The neuroblastic zone, in which rods are presumed to originate, is located in the outer half of the inner nuclear layer at this stage (Kuwabara and Weidman, 1974). Therefore, we can assume that the young rods must migrate across the developing outer plexiform layer to reach the outer nuclear layer, although this process has not been specifically studied in mammals. It is quite possible, however, that the same phenomenon that I have described in the fish retina also occurs in the developing mammalian retina, namely, radial fibers of Müller cells may guide vertically-migrating rod neuroblasts to their proper laminar destination. This raises the possiblity that Müller cells might play an important role in the histogenetic mechanisms responsible for the formation of cytoarchitectonic layers in the vertebrate retina.

ACKNOWLEDGEMENTS

This work was supported by NIH grant EY-04318 and the Alfred P. Sloan Foundation. The author has published previously as P. R. Johns.

REFERENCES

Blaxter, J. H. S. and M. P. Jones. 1967. The development of the retina and retinomotor responses in the herring. J. Mar. Biol. Assoc. U.K. 47:677-697.

Blaxter, J. H. S. and M. Staines. 1970. Pure-cone retinae and retinomotor responses in larval teleosts. J. Mar. Biol. Assoc. U.K. 50:449-460.

Boulder Committee. 1970. Embryonic vertebrate central nervous system: Revised terminology. Anat. Rec. 166:257-262.

Carter-Dawson, L. D. and M. M. LaVail. 1979. Rods and cones in the mouse retina. II. Autoradiographic analysis of cell generation using tritiated thymidine. J. Comp. Neurol. 188:263-272.

Cooper, N. G. F. and D. A. Steindler. 1986. Lectins demarcate the barrel subfield in the somatosensory cortex of the early postnatal mouse. J. Comp. Neurol. 249:157-169.

Hatten, M. E., R. K. H. Liem and C. A. Mason. 1984. Two forms of cerebellar glial cells interact differently with neurons in vitro. J. Cell Biol. 98:193-204.

Johns, P. R. and R. D. Fernald. 1981. Genesis of rods in teleost retina. Nature 293:141-142.

Johns, P. R. 1982. Formation of photoreceptors in larval and adult goldfish. J. Neurosci. 2:178-198.

Kaplowitz, P. B. and A. A. Moscona. 1976. Stimulation of DNA synthesis by ouabain and concanavalin A in cultures of embryonic neural retina cells. Cell Different. 5:109-119.

Kuwabara, T. and T. A. Weidman. 1974. Development of the prenatal rat retina. Invest. Ophthalmol. 13:725-739.

Mann, I. C. 1928. The process of differentiation of the retinal layers in vertebrates. Brit. J. Ophthalmol. 12:449-478.

Meshul, C. K., F. J. Seil, and R. M. Herndon. 1987. Astrocytes play a role in regulation of synaptic density. Brain Res. 402:139-145.

Nork, T. M., I. H. L. Wallow, S. J. Sramek, and G. L. Anderson. 1987. Müller cell involvement in proliferative diabetic retinopathy. Invest. Ophthalmol. Vis. Sci., Suppl. 28:123.

Price, J., D. Turner and C. Cepko. 1987. Lineage analysis in the vertebrate nervous system by retrovirus-mediated gene transfer. Proc. Natl. Acad. Sci. USA 84:156-160.

Ramon y Cajal, S. 1892. The Structure of the Retina. translated by S. A. Thorpe and M. Glickstein, Charles C. Thomas, Springfield, IL, 1972, 196 pp.

Ramon y Cajal, S. 1911. Histologie du Systeme Nerveaux de l'Homme et des Vertebres, Maloine, Paris. Reprinted by Consejo Superior de Investigaciones Cientificas, Madrid, 1955, 993 pp.

Rakic, P. 1971. Neuron-glia relationship during granule cell migration in developing cerebellar cortex. A Golgi and electron microscopic study in Macacus rhesus. J. Comp. Neurol. 141:283-312.

Rakic, P. 1972. Mode of cell migration to the superficial layers of fetal monkey neocortex. J. Comp. Neurol. 145:61-84.

Raymond, P. A. 1985. The unique origin of rod photoreceptors in the teleost retina. Trends in Neurosci. 8:12-17.

Raymond, P. A. and P. K. Rivlin. 1987. Germinal cells in the goldfish retina that produce rod photoreceptors. Develop. Bio. 122:120-138.

Sandy, J. M. and J. H. S. Blaxter. 1980. A study of retinal development in larval herring and sole. J. Mar. Biol. Assoc. U.K. 60:59-71.

Sheffield, J. B. and H. P. Li. 1987. Interactions among cells of the developing neural retina in vitro. Amer. Zool. 27:145-159.

Sidman, R. L. and P. Rakic. 1973. Neuronal migration, with special reference to developing human brain: A review. Brain Res. 62:1-35.

Silver, J., S. E. Lorenz, D. Wahlsten and J. Coughlin. 1982. Axonal guidance during development of the great cerebral commissures: Descriptive and experimental studies, in vivo, on the role of preformed glial pathways. J. Comp. Neurol. 210:10-29.

Smith, G. M., R. H. Miller and J. Silver. 1986. Changing role of forebrain astrocytes during development, regenerative failure, and induced regeneration upon transplantation. J. Comp. Neurol. 251:23-43.

Turner, D. L. and C. L. Cepko. 1987. A common progenitor for neurons and glia persists in rat retina late in development. Nature 328:131-136.

Uga, S. and G. K. Smelser. 1973. Electron microscope study of the development of retinal Müllerian cells. Invest. Ophthalmol. 12:295-307.

Wilson, C. A., E. Stefansson, T. Schoen and T. Kuwabara. 1987. Mitosis in the adult rat retina after temporary interruption of retinal blood flow. Invest. Ophthalmol. Vis. Sci., Suppl. 28:107.

Plasticity of Retinal Glioblast Cells: Neuronal Contact Regulates Phenotypic Maturation of Embryonic Müller Cells

Paul J. Linser

The archetypical radial glial cell, the retinal Müller cell has been considered to be a "primitive" form of glia (Cajal, 1892). Nevertheless the Müller cell is a highly specialized cell type whose form and function have been apparently conserved throughout vertebrate evolution. All sighted vertebrates ranging from cyclostomes to man possess Müller cells with strikingly similar structure and common biochemical characteristics (Hueter et al, 1986; Linser and Moscona, 1984a). Discovering the roles performed by these glial cells will lead to understanding of fundamental principles in retina function. Also, since it is widely believed that radial glia play formative roles in the development of form and function throughout the vertebrate nervous system, developmental analyses of Müller cells will also aid our understanding of neural development in general.

Our laboratory has been engaged in studies of Müller cell development and the dynamic relationship between these cells and other embryonic retinal cells. In order to follow the development of any cell type it is necessary to first establish characteristics which can be used to identify the cell. Numerous biochemical markers have been described which are characteristic of mature Müller cells in many animal species (Eisenfeld et al, 1985; Kumpulainen et al, 1983; Linser, 1987; Lemmon, 1986; Lemmon and Reiser, 1983; Terenghi et al, 1983). Among several characteristics which have been found in a broad range of vertebrate Müller cells, two enzymes have been shown to be characteristic of Müller cells in at least one example of every extant class of vertebrates. These are the enzymes glutamine synthetase (GS) and carbonic anhydrase-II (CA-II) (Hueter et al, 1986; Linser and Moscona, 1984a). In all species so far examined, immunochemical analyses of GS can be used to distinguish Müller cells from retinal neurons. CA-II is also found ubiquitously in Müller cells but certain animal species also possess this enzyme in limited subcatagories of retinal neurons (Hueter et al, 1986; Linser et al, 1985). In the mature chick retina, the model with which our laboratory has performed most of its developmental analyses, GS and CA-II are immunodetectable only in Müller cells.

The patterns of expression of these two enzymes during chick retina development contrast dramatically. In some ways, the expression patterns of these two enzymes exemplify the two general patterns of expression that we and others have observed in Müller cell development.

Specifically, biochemical markers particular to Müller cells in mature retina are either expressed late in development (i.e. after neuronal maturation has begun) and only in Müller cells or are expressed very early in development (i.e. prior to the appearance of distinct neurons) in all cells and only later restricted to the glia as neurons differentiate and cease to express the marker.

GS AND CA-II EXPRESSION

The expression of GS is a specific characteristic of functionally mature Müller cells. In the chick retina, GS levels are extremely low until the retina has achieved structural and to some extent functional maturity. Under normal circumstances, GS expression in Müller cells begins suddenly at day 16 of development (reviewed in Linser and Moscona, 1984b and figure 1). The sudden expression of this enzyme is induced by systemic corticosteroids and the level of GS in the tissue increases more than 100 fold between day 16 and hatching 5 days later (Moscona and Linser, 1983). GS is a central enzyme in the turnover of neuroactive amino acids such as glutamate and GABA (Hamberger et al, 1979; Linser, 1987). The need for high levels of GS is presumably not great until neuronal activity commences. Hence GS expression is not triggered until late in retina development. Although corticosteroid induction may be most pronounced in chick or precocial birds, late expression of GS has also been demonstrated in mammalian retina (Linser et al, 1984). Müller cells capable of expressing GS in response to artificially applied corticosteroids can be detected in the chick retina several days before normal expression commences. As early as day 7-8 of development, a few radial Müller glial cells can be identified in the fundal region of retinas stimulated precociously with hormone (reviewed in Moscona and Linser, 1983). Even at this time however, certain retinal neurons have already differentiated days earlier (Kahn, 1974).

The pattern of expression of CA-II is quite different (Figure 1). Using polyclonal and monoclonal antibodies we have been able to define four phases in the expression of CA-II in the embryonic chick retina. The first phase, or the preinduction phase, is the only period of ocular development during which neural ectoderm of the eye shows no detectable CA-II. This is the period from the beginning of development until 19–20 somite pairs in the embryo.

The next phase, or the primary induction phase, corresponds to the time in development when noticeable changes in tissue architecture of the eye primordium first appear. At 20 somite pairs, induction of the lens placode and optic cup are first indicated by changes in cell orientation and shape (Coulumbre,

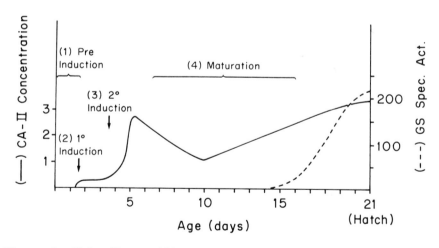

Figure 1. This figure illustrates the development of the chick retina with regard to the normal appearance of glutamine synthetase (GS) and 4 phases in the expression of carbonic anhydrase-II (CA-II). GS levels are expressed in specific activity units (Linser and Moscona, 1984b). CA-II levels are shown as concentration with the units in µg/mg total protein (days 3 through hatching, Linser and Moscona, 1981) or in arbitrary units based on immunohistochemistry (days 0 through 3).

1961). At this time CA-II becomes detectable in a few cells of both the optic neural ectoderm and the overlying skin ectoderm. During the period from 20 somite pairs (i.e apx. 36hr.) until roughly day 3, CA-II remains immunodetectable in the optic cup but levels are low.

Between day 3 and 4 of development, the choroid fissure closes and the third phase of retinal CA-II expression begins. At this time the concentration of CA-II in the presumptive retina increases dramatically such that by day 5 of development CA-II represents 3% of total retinal protein (Linser and Moscona, 1981). This phase can be called the secondary induction phase. During the second and third phases of CA-II expression all retinoblasts of the optic neural ectoderm can be immunolabelled for CA-II with monoclonal antibodies.

The final phase of CA-II expression commences with the appearance of the first differentiated cell type in the retina, the ganglion cell neurons, begining on day 5 (Kahn, 1974; Linser and Moscona, 1981). It is a general feature of neuronal maturation in the retina that as neurons become identifiable by virtue of position and morphology that they lose the capacity to be immunostained for CA-II (Linser and Moscona, 1981). As definitive neuronal cells emerge, Müller cells begin to appear and show enhanced capacity to immunostain for CA-II relative to staining intensities of individual cells earlier in development. This final phase of CA-II expression can be called the maturation phase.

During all of the phases of CA-II expression, the eye rudiment is distinguishable from most other embryonic tissues by virtue of CA-II immunolabelling capacity. Certain scattered mesenchymal cells in the head and blood cells also show CA-II early but the eye is quite distinct. Indeed, the border of the neural ectoderm which will form the optic stalk and later optic nerve is clearly distinguished from the presumptive retina tissue by a contrasting lack of CA-II labelling.

In our earliest published analyses of CA-II expression we failed to detect CA-II prior to day 3 of development (Linser and Moscona, 1981). Improved immunoreagents including a panel of monoclonal antibodies to CA-II have made it possible to investigate the details of the very early stages of retina development. Although numerous markers have been described which show early generalized expression followed later by restriction to Müller cells

(Lemmon, 1986; Lemmon and Reiser, 1983; Lemmon, personal communication; Linser, 1987; Linser and Perkins, 1987b), most studies have not reported investigations earlier than day 3 or 4 of development. Hence we cannot judge whether or not other markers show similar very early phases of expression comparable to CA-II.

NEURONAL-GLIAL INTERACTIONS AND MÜLLER CELL DEVELOPMENT

Having established GS and CA-II as markers of Müller cell maturation, I will now turn to studies of the role of cell interactions in Müller cell development. Studies performed by Moscona and his coworkers showed that GS expression is influenced by cell interactions. Specifically it was shown that dissociated embryonic retina cells could only mature and express GS if they were allowed to reassociate into multicelluar aggregates. Isolated retinoblasts of 6 day retina or older were incapable of developing the competence to express GS even in the presence of corticosteroids (Morris and Moscona, 1971; reviewed in Linser and Moscona, 1984b). Later, Linser and Moscona showed that the Müller cells were the specific cells which express GS and that their development to the point of GS expression was dependent on interactions between the glioblasts and retina neurons (Linser and Moscona, 1983). These studies focussed on the period of development which corresponds to the maturational phase of CA-II expression during which CA-II becomes glia specific. In contrast to GS, CA-II expression in Müller glioblast cells was not dependent on the same neuronal-glial interactions as those involved in regulation of GS (Linser and Moscona, 1983).

PROBES OF NEURONAL-GLIAL INTERACTIONS

The failure of GS expression in isolated retinoblast cells seemingly involves cell-cell contact between heterotypic cells, the neurons and the glia or glioblast cells (Linser, 1987). The use of cell surface binding probes provides the most direct evidence of contact dependence. In particular, plant lectins and antibodies have proven useful in blocking interactions which regulate GS expression competence in the glia.

LECTINS

Efforts to analyze cell interactions have focussed on cell surface ligands which might have the capacity to interfere with putative cell membrane components involved in heterotypic cell-cell communication. The succinylated derivative of the plant lectin conconavalin A (SConA) proved to be very useful in this respect. Specifcally we showed that SConA had the remarkable capacity to prevent neuronal glial interactions in reaggregating embryonic retina cells. Addition of SConA to dissociated cells placed in rotation culture did not reduce formation of multicellular aggregates. When the aggregates were then cultured for 7-10 days which normally results in high GS expression and retinotypic histogenesis, very little GS was produced (Linser, 1987). Furthermore, histological and immunochemical analyses showed that the neuronal and glial cells had sorted out from one another rather than interacting to form the normal glial-neuronal relationships (Linser, 1987). CA-II expression was nevertheless unperturbed in the glia even though GS was absent or dramatically reduced (Linser, 1987).

The inhibitory effects of SConA support the hypothesis that cell-cell contact is required for GS expression. Indeed, the aggregates formed in the presence of the lectin contain both neurons and glia within the same tissue mass but the heterotypic cells are spatially segregated from one another. Thus, GS expression apparently requires not only close proximity of neurons and glia but also direct contact.

Additional insights were gained from the analyses with SConA. First of all, full inhibitory effect of the lectin required the presence of the lectin only during the initial 24hrs of the 7-10 day culturing period (Linser, 1987). Therefore, it is evident that formation of developmentally important neuronal-glial associations in vitro involves early events and reflects a transient capacity of the cells. Other investigations support the transient capacity of embryonic retina cells to reassociate in a tissue-typic manner. For example, glial-derived retina cells can be readily prepared to high purity by culturing dissociated retina in monolayer in the presence of neurotoxic chemicals (Moscona et al, 1983) or neurolytic antibodies (Linser and Moscona, 1983). This type of procedure, however, generally requires one or more days in culture. The surviving glial cells express CA-II and little or no GS as expected. The glia purified in this way become completely refractile to reassociation with neurons and hence to any efforts to stimulate GS expression (Moscona et al, 1983; Linser, unpublished observations). Indeed, when glia prepared in this way are reaggregated with neurons the two populations of cells sort out into spatially separated areas of the aggregates which form (Moscona et al, 1983). The resultant aggregates look very much like those generated with SConA in that the glia always take an internal position in the aggregate surrounded by an enveloping layer of neuronal cells (Linser, 1987; Moscona et al, 1983).

These and other results indicate that retina glioblast cells rapidly lose their affinity for neuronal association when separated from the neuronal cells. Hence, in addition to losing the capacity to express GS, glia seem to lose other qualities as a result of being deprived of neuronal contact. Also, these changes apparently take place quite rapidly.

SConA also allowed us to analyze membrane components which might be involved in neuronal-glial interactions. Isolated plasma membrane material was analyzed by SDS-PAGE, Western blotting and then lectin labelling of the blot (Linser, 1987). We compared membrane material from aggregation competent 7 day retina with that of a later embryonic age (15 day) which had lost most of its reaggregation competence (Linser, 1987; Morris and Moscona, 1971). We also compared membrane material from trypsin dissociated and recovered 7 and 15 day retina cells. The plasma membrane material isolated from fresh tissue showed few remarkable differences with regard to SConA-binding components (Linser, 1987). However, when we examined the membrane material from trypsinized and recovered cells, we found that the younger retina responded to the challenge of trypsinization by producing very high levels of certain SConA binding components whereas the older retina cells produced a pattern very similar to the intact tissues (Linser, 1987).

These results suggest that early, reaggregation competent retina cells possess the capacity to elaborate increased levels of certain plasma membrane glycoconjugates when challenged with conditions requiring reassociation. Older retina cells which do not aggregate well (Morris and Moscona, 1971) do not respond in this way. Hence, some of the components recognized by the lectin may be directly responsible for cell reassociation. This is supported by the, observation that lentil lectin, which binds a subset of ConA-binding glycoconjugates (Kornfeld et al, 1981) does not mimic the effects of SConA on cell reassociation. Lentil lectin also does not identify the SConA-binding membrane components which show enhanced levels following trypsin-recovery of young retina cells (Linser, 1987 and unpublished observations).

ANTIBODIES

Antibody probes have also provided support for the importance of cell contact in GS regulation. Early experiments showed that purified IgG from rabbit antiserum raised against total retina cell surface could reduce GS expression in vitro (Linser, 1987). More recently we have employed monoclonal antibodes to study cell surface components more closely. In our efforts we have generated several monoclonal antibodies to retina cell surface molecules ranging from gangliosides to the adhesion molecule N-CAM. One particular monoclonal antibody called 5A11 has proven to be of particular interest in our studies of Müller cell development.

The 5A11 antibody was generated by immunizing mice with trypsin dissociated and recovered 7 day embryonic retina cells. The hybridoma cell line was selected by sequential screens including ELISA assay of binding to membrane vesicles and immunohistochemical staining of frozen retina sections. The antigen recognized by the 5A11 antibody is developmentally regulated and exhibits a changing pattern of distribution during development.

At very early stages of development, the 5A11 antibody immunostains much of the embryo. Figure 3 shows that at three days of development the early retina, lens and brain stain well and the head mesenchyme also shows significant labelling. In comparison, CA-II antibodies label the ectodermal portions of the eye and certain scattered cells in the head mesenchyme. Two days later however, the 5A11 staining in tissues outside the eye has decreased dramatically and the presumptive retinal pigment epithelium (RPE) and retina show most intense labelling. Significant staining is still evident in the brain and blood cells at this time. This overall pattern of very intense labelling in retina and RPE with lower yet significant staining of brain and blood cells persists through maturity (Linser and Perkins, 1987b). During the early stages of retina development, the 5A11 antigen is seemingly present on all retinoblast cells. Later as specific cell type differentiation commences, the antigen distribution shifts such that neuronal cells seem to lose the capacity to immunostain for 5A11 (Figure 2). Monolayer cultures of embryonic retina (or brain) cells show surface labelling of the glial-derived flat cells and not the round neuronal cells. Also, double labelling for CA-II and 5A11 show virtually identical patterns of labelling throughout retina development. As with CA-II, 5A11 becomes gradually localized predominantly to Müller cells (Linser and Perkins, 1987b). The relative concentration of 5A11 antigen in retina increases at least 6-fold between days 6 and 15 of development. At hatching, double-label analyses show that 5A11, GS and CA-II colocalize in retina sections (Figure 3). In addition to the Müller cells, the 5A11 antigen is also present on the surface of photoreceptor cells and on or between the outer segments of the rods and cones (Figure 3 and Linser and Perkins, 1987b).

The 5A11 antigen is a membrane component and cannot be recovered from water soluble fractions of retina lysates. When retina tissue is dissociated with trypsin or papain, double-labelling for GS and 5A11 shows that Müller cells retain the antigen (Figure 4). The antigen can be solublized with certain nonionic detergents and is destroyed by protease-K (Linser, unpublished observations). Thus it appears that the 5A11 antigen is probably an integral membrane protein of mature Müller cells and to a lesser extent photoreceptors. Efforts to characterize the antigen have been slowed by the fact that antigenicity is very rapidly and irreversably lost in the presence of SDS thus making Western blot analyses difficult. Very recent efforts utilizing ion exchange chromatography, immunoprecipitation and acetone precipitation have implicated a protein with an apparent molecular weight in SDS-PAGE of 40K daltons as the 5A11 antigen. This conclusion is tentative and will require further analyses.

The gradual shift from generalized labelling of retinoblasts to predominantly Müller cell localization places the 5A11 antigen into the catagory of Müller cell markers which includes CA-II as described earlier. Our goal in generating this antibody was not to produce additional markers of Müller cell maturation. Our intention was to produce probes of the cell surface which could interfere with neuronal-glial interactions and the maturation of the Müller cells. In collaboration with Dr. Rob Hausman we have found

38

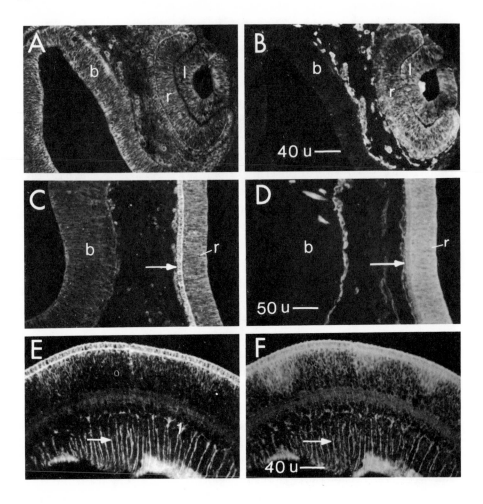

Figure 2. Double-label immunohistochemical comparison of the distributions of the 5A11 antigen (A,C,E) and CA-II (B,D,F) in 3 (A,B), 5(C,D) and 11(E,F) day embryos and retinas. Frozen sections were simultaneously labelled with monoclonal 5A11 and polyclonal CA-II antibodies by standard techniques (Linser,1987). At three days, 5A11 staining (A) is present in brain (b), retina (r) and lens (l) and surrounding cells of the head, whereas CA-II (B) is primarily localized to the eye primordium including lens and retina with the brain negative. At 5 days (C,D) retina (r) shows enhanced labelling for 5A11 (C) and CA-II (D). The brain (b) shows significant 5A11 but no CA-II. The retinal pigmented epithelium (arrows) shows particularly intense labelling for 5A11. By day 11 (E,F), staining of retina shows that 5A11 (E) and CA-II (F) both define the emerging Müller cells as indicated by the Müller cell processes in the nerve fiber layer (arrows).

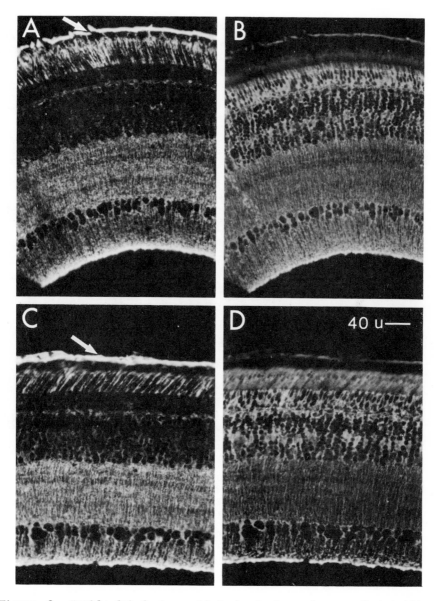

Figure 3. Double-label immunohistochemical analyses of hatchling retinas labelled for 5A11 and GS (A,B) or 5A11 and CA-II (C,D) respectively. Note similarities of staining patterns. Also note intense labelling for 5A11 in the pigmented epithelium (arrows) and to a lesser extent in the photoreceptor process layer.

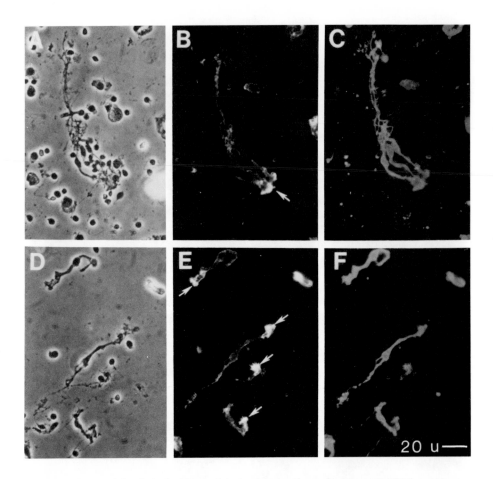

Figure 4. Double-label immunohistochemical analysis of 5A11 antigen (B,E) and GS (C,F) in isolated cells prepared from papain-dissociated hatchling retina. Panels A and D are phase contrast images of the two fields shown. Note complete correspondence between 5A11 and GS positive cells. Also note that the apical microvillar crowns of the identified Müller cells (arrows) show particularly intense 5A11 labelling indicating non-uniform distribution of the antigen on the Müller cell surface. Photoreceptors, both rods and cones (not shown) also label with 5A11 but to a lesser intensity than Müller cells.

that the antibody can reduce retina membrane vesicle aggregation <u>in</u> <u>vitro</u> (personal communication). Furthermore, we have found that the 5A11 antibody can also inhibit GS production in tissue culture.

Figure 5 shows two experiments in which either concentrated 5A11 hybridoma supernatant (A) or purified 5A11 antibody (B) were

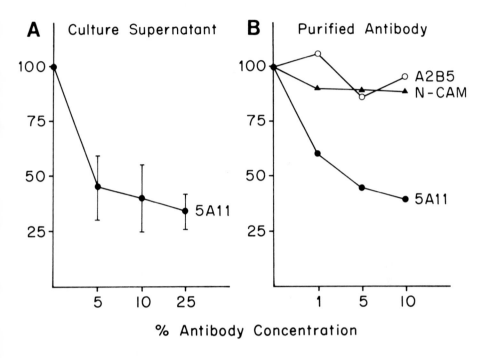

Figure 5. This pair of graphs shows the inhibitory effects of 5A11 antibody on the expression of GS in tissue cultures of isolated embryonic retina cells. 7 day embryonic retinas were dissociated with trypsin and plated in monolayer culture at moderately high cell densities to promote cell interactions. 5A11 antibody was included at various concentrations. In (A) 10X-concentrated 5A11 hybridoma culture supernatant was added and in (B), purified antibody was used. Also in (B), parallel cultures received purified antibodies to N-CAM or the neuron-specific ganglioside recognized by the A2B5 antibody. In all cases, the GS activity was measured and is plotted relative to untreated control cultures whose activity was set to 100 percent. Error bars in (A) represent +/- 1 SD. Abscissa: Added antibody concentration (%). Ordinate: Relative GS activity.

added to retina cells in culture. In both cases the antibody showed the capacity to reduce GS expression in a dose-dependent manner. Control antibodies to other cell surface antigens showed no significant effect in comparison.

To summarize our knowledge of 5A11, the antigen is apparently a membrane protein which is developmentally regulated in retina, appearing early on all retinoblasts and later becoming restricted to the Müller cells and to a lesser extent the photoreceptor cells. As with other Müller cell markers which are expressed early in embryonic retina, most neurons can be readily distinguished in developing retina as they emerge in identifiable form by their lack of 5A11 reactivity. In contrast, morphologically distinct embryonic Müller cells show enhanced immunolabelling for 5A11 relative to earlier antigen-positive blast cells again as the Müller cells become distinguishable. Finally, the 5A11 antibody can interfere with membrane vesicle adhesion and also Müller cell maturation as defined by GS expression. It has long been noted that the capacity of dissociated embryonic retina cells to aggregate and undergo in vitro development including the expression of GS declines with developmental age of the starting tissue (Morris and Moscona, 1971). It is attractive to suggest that this developmental decline may reflect such changes in cell membranes as the loss of 5A11 antigen from retina neurons. Steinberg has shown that when cells of differing adhesivities are coaggergated, the more adhesive cells will always come to occupy the central core of the resultant aggregate (Steinberg, 1964). Perhaps the central clustering of glial derived cells in SConA aggregates and aggregates of precultured glia and neurons reflects similar adhesive differences mediated in part by the presence of 5A11 on glia and its absence from neurons. These are clearly speculative comments but represent testable hypotheses for future study.

CONCLUSIONS AND COMMENT

Perhaps the most striking characteristic of Müller cells is their capacity for change in response to altered relationships between the glia and neurons. I have discussed the regulation of GS by neuronal contact. In this case it seems reasonable to argue that the glia are "instructed" by neuronal cells to produce an enzyme in quantities necessary to satisfy the needs of the neurons. Communication of this need seems to be through contact. In the absence of contact-communicated "instructions" from the neurons, Müller cells exhibit plasticity and the capacity to change. I have also discussed apparent changes in cell membrane composition which lead to altered affinities between neurons and glia.

Others have shown that embryonic Müller cells will actually begin to express new phenotypic characteristics after removal of neuronal influence. Specifically, isolation of Müller cells from neuronal contact can lead to the expression of lens-like characteristics including production of crystallins and the lens membrane protein MIP26 (Moscona et al, 1983). This phenotypic shift has been viewed as a conversion from one tissue cell type (neural retina Müller cell) to another (skin ectoderm lens) (Moscona et al, 1983). We have recently described the transient presence of a subcategory of Müller cell that we called retina-optic nerve boundary cells which express delta crystallin during normal development (Linser and Irvin, 1987). Hence the expression of lens characteristics may represent a normal potential in the differentiation repertory of Müller cells. This potential seems to be under the direct influence of neuronal cueing.

Recently, in colaboration with G. Engbretson, we have found that the glial cells of the reptillian parietal eye share characteristics with retinal Müller cells including GS (unpublished observations). In this system a visual apparatus develops from an expansion of the diencephalon as with the "normal" eye. In the case of the parietal eye however, the neural ectoderm never infolds on itself so that a double epithelium does not form (Eakins, 1973). The result is that the front portion of the optic vesicle differentiates into a lens and the posterior into retina with photoreceptor cells, ganglion neurons and glial cells (Eakins, 1973). The glia have similar structure to Müller cells and contain GS (unpublished observations). It is interesting to speculate that a factor which leads to lens differentiation in the anterior neural ectoderm and glial differentiation in the posterior is the absence or presence of neurons respectively.

Perhaps from this model system and the expression of lens characteristics by Müller cells in the absence of neurons in culture one can infer that lens phenotype is a normal program of differentiation in optic ectoderm which can be suppressed by neuronal contact. In the absence of specific neuronal-glial contact communication, the Müller cell derivatives "default" to the lens-like phenotype. In the same vein then Müller cell differentiation is intimately tied to and perhaps directed by the differentiating neurons with which they are in contact.

These comments clearly contain speculation. The data do strongly support the notion that the differentiative state of Müller cells is very plastic. The functional dichotomy of neurons and glia

44

seen in the mature retina is typical of all complex neural tissue. Recent studies have shown similar plasticity and neuronal control over the development of neuroglial cells in other regions of the avian brain as well (Galileo and Linser, 1987; Linser and Perkins, 1987a). Hence it is logical to argue that the functional differentiation of Müller cells and glia in general is controlled by the immediate structural and physiological needs of their cellular environment. Our results show that the molecular basis of such developmental regulation is at least partially controlled through cell-cell contact interactions between the glia and their partners in this functional dichotomy, the neurons.

ACKNOWLEDGEMENTS.

This research supported by grant 1-1030 from the March of Dimes Birth Defects Foundation and grant BNS-8718080 from the National Science Foundation.

REFERENCES

Cajal, S. R. 1892, (1973). La retine des vertebres., La Cellule 9:119-225 Compiled and translated by S.A.Thorpe and M. Glickstein, In: The Vertebrate Retina,838-852.

Coulumbre, A. J. 1961. Cytology of the developing eye. Int. Rev. Cytol. 11:161-194.

Eakins, R. 1973. The Third Eye, Univ. Cal. Press, LA.

Eisenfeld, A. J, A. H. Bunt-Milan and J. C. Saari. 1985 Localization of retinoid-binding proteins in developing rat retina. Exp. Eye Res. 41:299-304.

Galileo, D. S. and P. J. Linser. 1987. In vitro development of immunomagnetically-purified A2B5(-) embryonic chick optic tectum cells. J. Cell Biol. 105:319a.

Hamberger, A. C., G. HanChiang, E. S. Nylen, S. W. Scheff and C. W. Cotman. 1979. Glutamate as a CNS transmitter. Evaluation of glucose and glutamine as precursors for the synthesis of preferentially released glutamate. Brain Res. 168:513-530.

Hueter, R. E., J. A. Moffat, B. Battelle and P. J. Linser. 1986. Immunochemical analysis of glutamine synthetase and carbonic anhydrase in the retina of a cyclostome. Invest Ophthal. Vis. Sci. (supplement). 27:230.

Kahn, A. J. 1974. An autoradiographic analysis of the time of appearance of neurons in the developing chick neural retina. Develop. Biol. 38:30-40.

Kornfeld, K., M. L. Reitman and R. Kornfeld. 1981. The carbohydrate-binding specificity of Pea and Lentil lectins. Fucose is an important determinant. J. Biol. Chem. 256:6633-6640.

Kumpulainen, T., D. Dahl, L. K. Korhonen and S. H. M. Nystrom. 1983. Immunolabelling of carbonic anhydrase isoenzyme C and glial fibrilary acidic protein in paraffin-embedded tissue sections of human brain and retina. J. Histochem. Cytochem. 31:879-886.

Lemmon, V. 1986. Localization of a Filamin-like protein in glia of the chick central nervous system. J. Neurosci.6:43-51.

Lemmon, V. and G. Reiser. 1983. The developmental distribution of vimentin in the chick retina. Dev. Brain Res. 11:191-197.

Linser, P. J. 1987. Neuronal-glial interactions in retina development. Am. Zool. 27:161-169.

Linser, P. J. and C. K. Irvin. 1987. Immunohistochemical characterizatioin of delta crystallin-containing retina/optic nerve "boundary" cells in the chick embryo. Develop. Biol. 121:499-509.

Linser, P. J. and M. Perkins. 1987a. Gliogenesis in the embryonic avian optic tectum: Neuronal-glial interactions influence astroglial phenotype maturation. Develop. Brain Res. 31:277-290.

Linser, P. J. and M. S. Perkins. 1987b. Regulatory aspects of the in vitro development of retinal Müller glial cells. Cell Diff. 20:189-196.

Linser, P. and A. A. Moscona. 1984a. Variable CA-II compartmentalization in vertebrate retina. Ann. NY Acad. Sci. 429:430-446.

Linser, P. and A. A. Moscona. 1984b. The influence of neuronal-glial interactions on glial-specific gene expression in embryonic retina. In: Gene Expression and Cell-cell Interactions in the Developing Nervous System. J. M. Lauder and P. G. Nelson eds, Plenum Press, N.Y. 185-202.

Linser, P. and A. A. Moscona. 1979. Induction of glutamine synthetase in embryonic neural retina: localization in Müller fibers and dependence on cell interactions. Proc. Natl. Acad. Sci. USA. 76:6476-6480.

Linser, P. and A. A. Moscona. 1981. Carbonic anhydrase-C in the neural retina: Transition from generalized to glia-specific cell localization during embryonic development. Proc. Natl. Acad. Sci. USA. 78:7190-7194.

Linser, P. and A. A. Moscona. 1983. Hormonal induction of glutamine synthetase in cultures of embryonic retina cells: requirement for neuron-glia contact interactions. Develop. Biol. 96:529-534.

Linser, P. J., K. Smith and K. Angelides 1985. A comparative analysis of glial and neuronal markers in the retina of fish: variable character of horizontal cells. J. Comp. Neurol. 237:264-272.

Linser, P. J., M. Sorrentino and A. A. Moscona. 1984. Cellular compartmentalization of carbonic anhydrase-C and glutamine synthetase in developing and mature mouse neural retina. Develop. Brain Res. 13:65-73.

Morris, J. E. and A. A. Moscona. 1971. The induction of glutamine synthetase in aggregates of embryonic neural retina cells: correlations with differentiation and multicellular organization. Develop. Biol. 25:420-444

Moscona, A. A., M. Brown, L. Degenstein, L. Fox and B. M. Soh. 1983. Transformation of retinal glia cells into lens phenotype: expression of MP-26, a lens plasma membrane antigen. Proc. Natl. Acad. Sci. USA 80:7239-7243.

Moscona, A. A. and P. Linser. 1983. Developmental and experimental
    changes in retina glia cells: cell interactions and control of
    phenotype expressiion and stability. Curr. Top. Develop. Biol.
    18:155-188.
Steinberg, M. S. 1964. The problem of adhesive selectivity in
    cellular interactions, In: Cellular Membranes in Development.
    M. Locke (ed.) Academic Press, Inc., N.Y. 321-366.
Terenghi, G., D. Cocchia, F. Michetti, A. R. Diani, T. Peterson, D.
    F. Cole, S. R. Bloom and J. M. Polak, Localization of S-100
    protein in Müller cells of the retina-1. Light microscopical
    immunocytochemistry. Inv. Ophthal. Vis. Sci. 24:976-980.

Positional Cues in the Developing Eyebud of *Xenopus*
Nancy A. O'Rourke and Scott E. Fraser

The formation of spatial patterns in the nervous system appears
to require a complex series of interactions. In the visual system,
this not only involves the patterning of neurons and glia within the
eye, but also the patterning of the spatially ordered optic
projections in visual centers of the brain. An ideal system for
studying patterning in the optic projections is the retinotectal
projection which forms the main visual pathway in lower vertebrates.
In the retinotectal projection, the retinal ganglion cells in the
eye project along the optic nerve and into the midbrain where their
connections form a topographic map of the retina over the surface of
the contralateral optic tectum. In this well ordered pattern 1)
neurons in a particular part of the retina consistently project to a
particular part of the tectum, and 2) neurons in neighboring
positions in the retina project to neighboring positions in the
tectum giving the projection a smooth internal order. Lower
vertebrates are capable of regenerating their optic nerve after
injury and eventually reforming this same pattern of connections. A
multitude of studies has examined the ability of optic nerve fibers
to consistently find their correct target region in the tectum both
during development and regeneration. Evidence from these studies
supports the existence of positional cues in both the retina and the
tectum which guide the fibers to their proper targets in the tectum.
Our current approach has focused on the role of positional cues in
the initial formation of the topographic map in **Xenopus laevis**,
commonly known as the South African clawed frog.

In the regenerating retinotectal system, evidence for
positional cues comes from studies in which small pieces of the
tectum were surgically excised and then reimplanted in a different

position or orientation. In the majority of cases, regenerating fibers were able to locate and terminate on the proper, though mispositioned, piece of the tectum (c.f. Yoon, 1975). In a related set of experiments in the newt (Fujisawa et al, 1981, 1982) and the goldfish (Meyer, 1984), regenerating fibers which entered the tectum in unfamiliar regions, were found to take tortuous pathways to reach their proper tectal target site. The results of these studies suggest that the optic fibers possess positional information which allows them to discriminate between positional cues in the tectum and use these cues to find their correct tectal targets. The evidence that positional cues exist in the regenerating system supports the idea that they could be present during initial development of the projection as well.

An important approach to testing for the presence of positional cues, often termed "positional information" (Wolpert, 1969, 1971), in the developing eyebud has been the study of pattern regulation. Pattern regulation refers to the changes in the patterns of positional information which can occur following the removal or rearrangement of embryonic tissue. In some cases, embryonic tissues respond to surgical manipulations by replacing the missing elements or forming anomalous structures such as duplicated pattern elements. In one study in **Xenopus**, wedges of the eyebud were grafted into heterotopic sites in the opposite pole of a host eyebud (Conway et al, 1980; Cooke & Gaze, 1983). The animals were then raised past metamorphosis and the patterns of the retinotectal projections were assayed electrophysiologically. In some animals, the descendants of the grafted cells projected to the tectum according to their original positions in the donor eyebud. This is known as a "mosaic" result. Because the cells project to their correct positions, this result suggests that the grafted cells possessed positional cues which were stable and passed on to their descendants. In other cases, the maps revealed a normal pattern in which the descendants of the grafted cells projected to the tectum according to their final positions in the retina. This "regulated" result suggests that the pattern of positional information in the grafted eyebud cells have been altered. The eventual understanding of the mechanisms involved in pattern alterations such as these should provide insight into the mechanisms by which the positional cues are initially provided to the cells in the embryo.

The studies of the heterotopic wedge grafts as well as other previous studies of pattern regulation in the eyebud suffer from a major limitation. In all such studies, the surgical manipulations were performed on the embryo, the animals were raised past metamorphosis, and then the retinotectal projection patterns were assayed electrophysiologically. In **Xenopus** as well as other lower vertebrates, the eye grows dramatically during the life of the animal by adding new cells at the ciliary margin around the circumference of the eye (Hollyfield, 1971; Straznicky and Gaze, 1971; Jacobson, 1976; Fernald, 1984). Cells are added in a radial pattern such that the descendants of any given cell are found in a radial wedge extending toward the margin of the adult eye. The retinotectal map assayed in a postmetamorphic animal is formed primarily by the descendants of the original embryonic cells which have been added to the eye during larval development. In the studies of pattern regulation in the eyebud, it is difficult to ascertain how the grafted cells have interacted to form the adult projection pattern. Consequently, such studies can reveal little about the mechanisms or timing of regulatory interactions.

To begin exploring the early events during regulatory interactions in the eyebud, we have developed a technique which allows us to label the early eyebud cells and assay their projection patterns in the live **Xenopus** larva. The technique was utilized to follow the initial development of topography in the normal retinotectal projection in vivo. It was then used in combination with experimental embryology to assay directly the presence and stability of positional cues in early eyebud cells and to evaluate the responses of grafted eyebud cells in regulatory interactions. Because the projection patterns of the grafted cells were traced in live animals, the same animals could be raised and the projection patterns of the descendants of the grafted cells could be assayed during late larval stages. This approach has revealed the timing of regulation and the identities of the cells contributing to the regulated patterns seen in postmetamorphic animals.

DEVELOPMENT OF RETINOTECTAL TOPOGRAPHY

The initial development of the retinotectal projection was followed using the in vivo fluorescent fiber-tracing technique (O'Rourke and Fraser, 1986a). The technique utilizes the fluorescent vital dyes, lysinated fluorescein dextran (LFD) and lysinated rhodamine dextran (LRD) which consist of 10,000 MW or larger dextran chains with fluorescent dyes and lysine molecules added along their lengths (Gimlich & Braun, 1985). In our studies, the dye was pressure injected into fertilized **Xenopus** eggs, labeling all the cells in the developing embryo. At later stages, labeled eyebud cells were grafted into the eyebuds of unlabeled host animals. The lysine allows the dyes to be fixed and the labeled cells can be visualized at later stages in frozen sections through the eye. The dye remains bright in cells of all layers of the retina and does not pass into the unlabeled cells of the host. In the center of the retina, where cell division ceases at early stages, the dye remains bright. Around the ciliary margin at the edge of the eye, where cell division continues at later stages, the dye becomes diluted. The technique, therefore, yields a group of prelabeled cells in the center of the retina. The dye remains diffusible in the cytoplasm of the cells and fills the axons of the developing neurons. Because the head of the early larvae becomes transparent, this permits the outgrowth of the fluorescently-labeled fibers to be visualized in the live tadpoles.

The vital-dye fiber-tracing technique was used to observe the initial appearance of topography in the projection. Fluorescently labeled eyebud fragments were grafted into equivalent positions in unlabeled host eyebuds at stage 30-32 (figure 1A). The grafted fragments from opposite poles of the eye were labeled with two different dyes (LFD and LRD). Since the dyes emit different wavelengths of light, the projections of the two populations of cells could be visualized and directly compared in the same host animal. The projection patterns from dorsal and ventral fibers were distinct as early as they were observed in the tectum at stage 40-41. The dorsal cells projected to the lateral tectum, while ventral cells projected to the medial tectum (figure 2A,B). In contrast, the nasotemporal topography appeared gradually. At stage 46, a stage when dorsoventral topography was already clear, the nasal and temporal fibers overlapped, projecting to the same area of the tectum (figure 3A,B). At stage 48-49, several days later, the nasal fibers had grown past the temporal fibers into more posterior regions of the tectum and the proper topographic order was apparent (figure 3C,D). Thus, along the dorsoventral axis the topography appears early, but along the anteroposterior axis topography appears

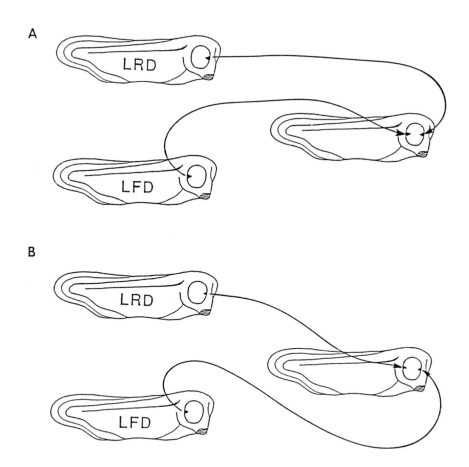

Figure 1. A) Homotopic fragment grafting operation. Two
fluorescently labeled donors were prepared, one labeled with
lysinated rhodamine dextran (LRD) and one labeled with lysinated
fluorescein dextran (LFD). A LRD labeled anterior eyebud fragment
was grafted into the anterior pole of the eyebud of an unlabeled
host. Similarly, a LFD labeled posterior fragment was grafted into
the posterior pole of the eyebud of the same unlabeled host. In
related experiments (not shown), a LRD labeled dorsal fragment and a
LFD labeled ventral fragment were both grafted into their positions
of origin in a single unlabeled host animal. The homotopic
operations were carried out on stage 30-32 animals. B) Heterotopic
fragment grafting operations. A LRD labeled anterior fragment was
grafted into the posterior pole of the eyebud of an unlabeled host.
A LFD labeled posterior fragment was grafted into the anterior pole
of the same unlabeled host. In a parallel set of experiments (not
shown), a LRD labeled dorsal fragment and a LFD labeled ventral
fragment were both grafted into the opposite poles of the eyebud of
a single unlabeled host. These operations were performed on animals
at stages ranging from stage 26 to 34.

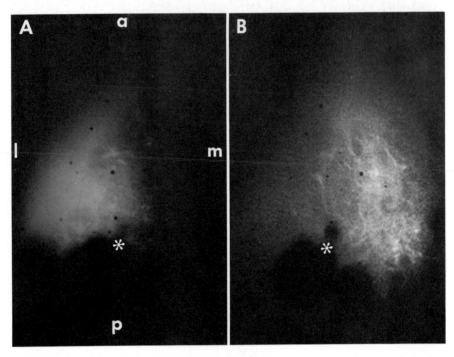

Figure 2. Tectal projection patterns of fluorescently labeled
dorsal and ventral optic fibers in a stage 47 host as seen from a
dorsal view. Epifluorescent images of the projection patterns in
the tectum were recorded by an image intensifying camera, enhanced
with an image processor, and displayed on a video screen. The
photographed areas show the fibers of passage entering the tectal
neuropil at the top of the panel and then forming arbors in the
tectal neuropil which fills most of the panel. The dark spots are
pigment cells in the pial membrane which can be used as landmarks
when comparing the panels. The asterisks mark the same pigment spot
in the two panels. A) The LRD labeled dorsal fibers project along
the lateral edge of the tectum. B) the LFD labeled ventral fibers
project more medially and form a projection pattern distinct from
that of the dorsal fibers. (l, lateral; m, medial; a, anterior; p,
posterior)

Figure 3. Tectal projection patterns of fluorescently labeled
anterior and posterior fibers in stage 47 and stage 49 hosts as seen
from a dorsal view. The images were obtained as described in figure
2. The asterisks mark the same pigment spots in adjacent panels.
At stage 46, the LRD labeled anterior fibers (A) and the LFD labeled
posterior fibers (B) project to roughly the same area of the tectal
neuropil and no obvious topography is seen. At stage 49, however,
the anterior fibers (C) have extended about 50 microns posteriorly
into new areas of the tectal neuropil, while the posterior fibers
have extended only a short distance (D). Thus, the topography along
the anteroposterior axis is set up gradually through the
preferential growth of the anterior fibers into posterior regions of
the tectum. (l, lateral; m, medial; a, anterior; p, posterior)

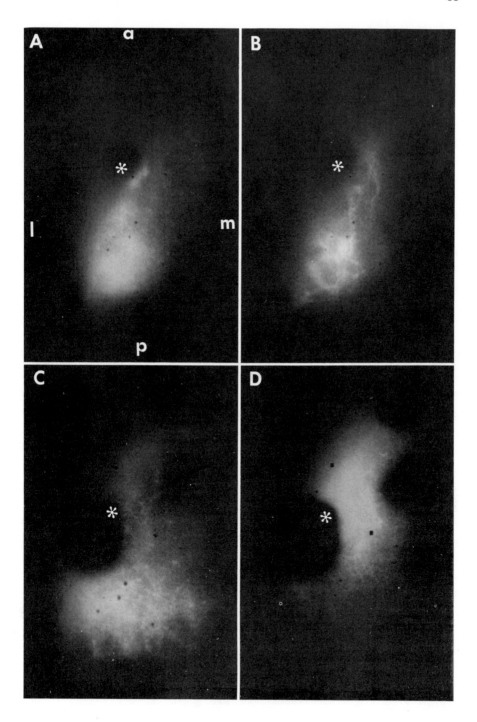

more gradually through a preferential growth of the nasal fibers into new territories of the tectum.

The finding that dorsoventral topography in the retinotectal projection was visible at the earliest stages we could visualize the fluorescent fibers in the tectum was in agreement with previous studies by Holt & Harris (1983). They used tritiated proline labeled dorsal and ventral eyebud fragments followed by autoradiography of the developing tectum to reveal that dorsal and ventral optic fibers projected to the tectum with normal topography upon first entering the tectum. The dorsal and ventral fibers were shown to project to the tectum in normal patterns even in the absence of neural activity, normal pathway of ingrowth (Harris, 1982, 1984) and the normal timing of ingrowth of the fibers (Holt, 1984). The results of these experiments suggest that these three factors are not dominant mechanisms involved in the formation of the retinotectal projection. This has led to the implication that positional cues could play a major role in initially guiding the dorsal and ventral optic fibers to their targets.

The fluorescent labeling technique permitted the cellular dynamics involved in the formation of anteroposterior topography to be followed directly. The projection patterns of groups of fibers could be visualized in the same animal over a period of days. In previous studies by Sakaguchi and Murphey (1985), the terminal arbors of small groups of fibers were visualized in wholemounts of the tectum using a cobalt staining technique. Terminal arbors during the early 40's stages were shown to have a much greater extent along the anteroposterior axis than along the dorsoventral axis, with single terminal arbors covering as much as 80% of the anteroposterior extent of the tectal neuropil during earlier stages. At later stages (stage 45), the percent tectal coverage was reduced as the tectal neuropil extended in length, and anteroposterior topography became apparent. Our in vivo studies demonstrated a similar early overlap but also permitted the same fibers to be followed as they sorted out into normal topographic order over a period of days. The dynamic refinement of topography along the anteroposterior axis suggests that there may be different mechanisms involved in formation of order along the two axes of the tectum. The in vivo fiber tracing methods are currently being used in our laboratory to further investigate the mechanisms involved in the initial development of topography in the retinotectal projection.

POSITIONAL CUES IN THE EYEBUD

Evidence from studies of the initial development of the retinotectal projection suggests that positional cues may play an important role in setting up the ordered projection. A more direct test of the existence of cell-autonomous positional markers in the early eyebud can be performed by challenging optic nerve fibers with inappropriate neighbors (Fraser, 1987). Very small groups of fluorescently labeled cells were grafted into an unlabeled host in heterotopic sites, for example, at the pole of the eyebud opposite to their position of origin (figure 1B). Unlike previous studies of heterotopic grafts, grafts containing as few as one retinal ganglion cell were grafted into a position in which they are completely surrounded by cells from the opposite pole of the eyebud. The fluorescent dextran labeling technique allowed the projection patterns of these grafted cells to be assayed directly as they first formed projections in the tectum. The use of LRD to label cells from one pole of the eyebud and LFD to label the cells from the opposite pole allowed easy evaluation of the topography of the initial projections. The results of these studies were very consistent. In all cases, the cells projected to the tectum according to their position of origin. Dorsal and ventral cells in heterotopic positions projected to the lateral and medial tectum respectively. Heterotopically grafted nasal and temporal cells first projected in overlapping territories of the tectum then sorted out into their normal topographic pattern at stage 48-49. These results indicate that the developing retinal ganglion cells possess positional information which guides their axons to their proper targets.

Results of previous studies in which topography was analyzed after metamorphosis have suggested that the stability of positional cues in the eyebud may depend on the experimental circumstances. For example, in some cases, the same surgical manipulations performed in different laboratories gave rise to distinctly different results. In a lab in which surgeries were performed in "full strength" embryonic media (approximately 50% Ringer's solution), the descendants of the grafted cells projected to the tectum in a manner appropriate for the position of origin of the graft (Gaze & Straznicky, 1980). That is, the grafted cells projected in a mosaic pattern suggesting that their positional cues

were stable. In another laboratory, the same grafts performed in
artificial pond water (about 10% Ringers), gave rise to regulated
projection patterns (Hunt & Jacobson, 1973). The regulated pattern
could be explained by an alteration of the positional cues in the
embryonic cells which was then passed on to their progeny. The
differences in the results of the two studies could be explained by
the use of different surgical media. The results of other studies
suggest that the stability of the positional cues in the cells may
depend on the developmental age of the cells. Surgical
manipulations performed at early stages gave rise to regulated
patterns while those performed at later stages gave rise to mosaic
patterns (Jacobson, 1968; Hunt & Jacobson, 1972). The results were
used to argue that "respecification" of the positional cues in the
eyebud cells can only take place at the earlier stages suggesting
that the positional cues are less stable at early stages.

To address some of these previous variabilities, the
heterotopic graft paradigm was used to test the stability of the
cell-autonomous positional cues in the early eyebud cells under a
variety of conditions. To test effects of grafting media on the
stability of positional cues, surgery was performed in: 1) full
strength media, 2) dilute artificial pond water, and 3) artificial
pond water intentionally titrated to a non-physiological pH (pH 6.5
or 8.0). Finally, to test the stability of positional cues in cells
of different ages, grafts were performed with host and/or donor
animals ranging from stages 26 to 34. In all cases, the retinal
ganglion cell axons projected to the tectum in a manner appropriate
to their position of origin in the donor animal. Repeated
observation showed that these projection patterns remained stable
until at least stage 52. Thus, it now appears that the differences
in the results of the earlier experiments cannot be explained by an
instability of the positional cues in the early eyebud cells which
is dependent on either surgical media or stage of the animals. In
the experiments described here, the positional cues in the cells
were stable under all circumstances tested.

STABILITY OF POSITIONAL CUES DURING PATTERN REGULATION

An understanding of the mechanisms involved in pattern
regulation in the eyebud promises to lead to a better understanding
of how positional cues are initially specified in the embryonic

tissue. To approach this issue we have tested the stability of the positional cues in grafted eyebud cells during a regulatory interaction which consistently leads to a regulated projection pattern when assayed in adult animals. The surgical manipulation involved replacing the temporal half of a right eyebud with the temporal half of a left eyebud, producing compound "nasal right-temporal left" or N_rT_l eyes. The half-eye replacement, which was performed at stage 32, requires an inversion of the dorsoventral axis of the grafted temporal left half-eyebud (figure 4A). In the original study of these compound eyes, the animals were then raised past metamorphosis and mapped electrophysiologically (Hunt & Jacobson, 1973). If the grafted cells projected to the tectum according to their original positions in the donor eyebud, the nasal visual field which is seen by the temporal half of the eye should have inverted dorsoventral polarity (figure 4C). Instead, maps of the N_rT_l eyes revealed either normal or double nasal patterns, both of which have normal dorsoventral polarity. In our own studies repeating the previous results in both postmetamorphic (figure 4D,E) (O'Rourke & Fraser, 1986b) and late larval animals (O'Rourke & Fraser, 1986c), the same patterns of regulated maps were obtained from all the N_rT_l eyes tested.

The results of the original N_rT_l study were interpreted as evidence for the respecification of the positional cues in the grafted temporal half of the eye. The descendants of the grafted cells would inherit the new pattern of positional values, resulting in the regulated pattern seen in the adult animal (Hunt & Jacobson, 1973). Other studies of pattern regulation in the **Xenopus** eyebud have led to the proposal of different mechanisms for pattern regulation which do not involve changes in the positional cues of the original grafted cells. In one proposed mechanism, regulation is thought to occur through a process of intercalary growth which is stimulated when normally nonadjacent cells are placed in neighboring positions in the eyebud (Cooke & Gaze, 1983). This growth gives rise to new cells with new patterns of positional values which fill in the missing pattern elements usually found between the confronted, but normally nonadjacent cells (French et al, 1976; Bryant et al, 1981). In another mechanism proposed by Ide and colleagues (1984, 1987), the cells are thought to change their positions within the eyebud during and after healing, leading to an apparent regulation. The regulated pattern seen in the maps of adult stage N_rT_1 eyes could arise through any of these mechanisms. An evaluation of the relative importance of the proposed regulatory mechanisms requires knowledge of the events which occur during early stages of pattern regulation.

The fluorescent fiber-tracing technique was used to investigate the stability of positional cues in the original grafted cells in the temporal half N_rT_1 of the eye during early larval stages (O'Rourke & Fraser, 1986b). At stage 24-26, LRD labeled dorsal left half-eyebud cells were grafted into the equivalent site in an LFD labeled host (figure 5). The resultant chimeric animals, with LRD labeled cells in their dorsal half-eyebud and LFD labeled cells in their ventral half, were allowed to heal about 18 hours until stage 32. The animals were then used as donors in N_rT_1 operations, resulting in a N_rT_1 eye in which the LRD labeled dorsal cells ended up in a ventral position and the LFD labeled ventral cells ended up in a dorsal position. This procedure permits both the position of the grafted cells and their projections to the tectum to be followed unambiguously. At stage 47, frozen sections through the temporal halves of the N_rT_1 eyes revealed that the cells remained in the positions in which they were grafted with virtually no mixing of cells. The dyes were diluted by the continued growth of the eye around the ciliary margin, indicating that the labeled cells

Figure 4. A) "Nasal right-Temporal left" or N_rT_1 grafting operation. The temporal half of a right eyebud is removed and replaced with the temporal half of a left eyebud at stage 32. The temporal left half is inverted along the dorsoventral axis during the operation. B) The normal pattern of an electrophysiological visuotectal map. C) Graphic representation of the expected "mosaic" pattern of the map of a N_rT_1 eye if the progeny of the grafted cells projected to the tectum according to the original positions of the grafted cells in the donor eyebud. Note that the nasal visual field, which is seen by the temporal half of the eye, has inverted dorsoventral polarity. D) Normal visuotectal map obtained from a N_rT_1 eye mapped in a postmetamorphic animal. E) Double-nasal map obtained from a N_rT_1 eye mapped in a postmetamorphic animal. In both D) and E), the nasal visual field which is seen by the temporal half of the eye maps to the tectum with normal dorsoventral polarity.

The data for the maps was collected by lowering a metal microelectrode into a set of positions in the tectum and finding the area of the visual field in which a light stimulus will elicit a response (receptive field) for each electrode position. The outline at the top represents the left tectum: c, caudal; r, rostral; m, medial; l, lateral. The circles represent the visual fields seen by each eye: d, dorsal; v, ventral; n, nasal; t, temporal. The numbers represent the electrode positions assayed in the tectum and the positions of the corresponding visual receptive fields.

59

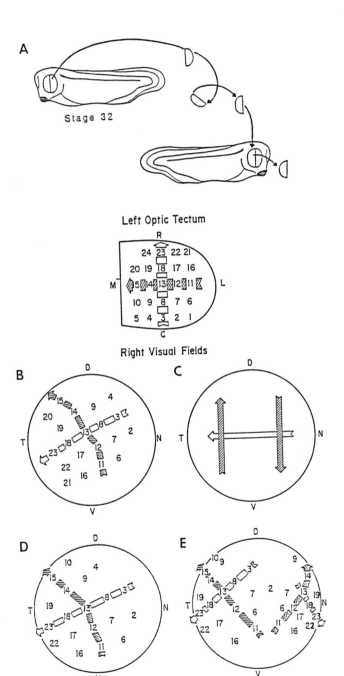

Left Optic Tectum

Right Visual Fields

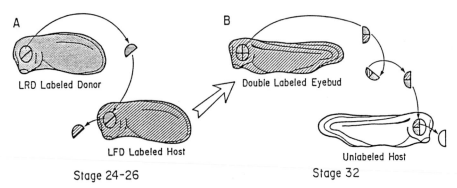

A — LRD Labeled Donor

LFD Labeled Host

Stage 24-26

B — Double Labeled Eyebud

Unlabeled Host

Stage 32

Figure 5. Fluorescent labeling of N_rT_1 eyes. A) A donor animal with a double labeled left eyebud was prepared by grafting the dorsal portion of the left eyebud of a LRD labeled animal into the dorsal part of the left eyebud of a LFD labeled animal at stage 24-26. The procedure results in an animal with a left eye which has a LRD labeled dorsal half and a LFD labeled ventral half. The border between the grafted and host halves is on the diagonal to compensate for asymmetric growth of optic stalk tissue into the eyebud which occurs between stage 26 and stage 32 (Holt, 1980). B) At stage 32 the animal with the double labeled left eye is used as a donor in a N_rT_1 operation. In this procedure, the LRD labeled dorsal cells end up in a ventral position and the LFD labeled ventral cells end up in a dorsal position.

were healthy and continued to contribute to the growth of the N_rT_1 eyes. Additionally, the labeled cells consistently projected to the tectum according to their position of origin in the donor eyebud and not according to the regulated pattern seen in the adult maps. Dorsal cells projected along the lateral edge of the tectum and ventral cells projected more medially (figure 6A,B). These results indicate that both the positions and the positional values of the grafted cells remain stable for some time and that early changes in the positional values of the original grafted cells do not contribute to the final regulated pattern.

GRADUAL APPEARANCE OF A REGULATED PROJECTION PATTERN

The stability of the positional information in the original grafted cells is in apparent contradiction to the regulated maps of N_rT_1 eyes in later stage larvae and adults. Since the topography of the N_rT_1 projections was assayed at early stages using the in vivo fluorescent labeling technique, the same animals could be raised and assayed at late tadpole or postmetamorphic frog stages. Again, the eyes consistently mapped in regulated patterns. The eyes in the

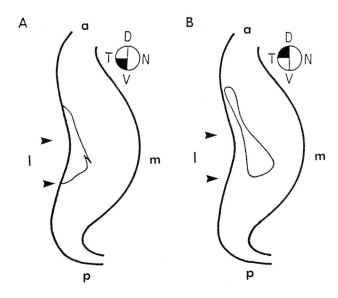

Figure 6. Graphic representation of the projection patterns from fluorescently labeled cells in the N_rT_1 eyes. The outer outline represents the border of the larval midbrain as seen from a dorsal view. (l, lateral; m, medial; r, rostral; c, caudal). The circles represent the N_rT_1 eyes with the final positions of the grafted cells shown in black. (d, dorsal; v, ventral; n, nasal; t, temporal). The fluorescently labeled fibers enter the contralateral midbrain at the anterior end, and project caudally along the optic tract and into the tectal neuropil, the anteroposterior extent of which is marked by the arrows. (A) The LRD labeled fibers project along the lateral edge of the tectum as if they were dorsal cells. (B) The LFD labeled fibers project medially and are not seen projecting along the lateral edge.

older animals, however, are comprised almost entirely of cells which have been added at the ciliary margin of the eye subsequent to stage 32. Studies using the interspecific **Xenopus borealis** cell marker (Thiebaud, 1983) in chimeric **laevis-borealis** N_rT_1 eyes show that the progeny of the grafted temporal cells make up roughly the same proportion of the adult eye as the grafted cells do in the embryonic eye (O'Gorman et al, 1987). It seemed certain that these newly added cells must contribute to the regulated pattern seen in the later stage compound eye maps. To test this, the projection patterns of the cells added to the eye subsequent to the embryonic stages were assayed using anterograde transport of horseradish peroxidase (HRP) (Fujisawa, 1984). A solution of HRP was carefully pressure injected into the eye through the iris. A tungsten needle was then used to injure small numbers of optic fibers around the limbus of the eye to obtain selective labeling of only the injured fibers. When the anterograde tracing technique was performed on normal eyes in mid-larval stage animals, the HRP labeled fibers from different parts of the eye were found to form clearly distinct projection patterns. For example, temporodorsal cells projected to the ventral (lateral as seen from a dorsal view) part of the tectum (figure 7A) and temporoventral cells projected to the dorsal (medial as seen from a dorsal view) part of the tectum (figure 7B), indicating that the HRP labeling technique provides a reliable method for assaying topography in the larval retinotectal projection.

The projection patterns of compound N_rT_1 eyes in midlarval stage **Xenopus** were assayed with the HRP technique to investigate the appearance of the regulated pattern (O'Rourke & Fraser, 1986c). HRP was introduced into small groups of fibers in the temporoventral quadrant of the N_rT_1 eyes. Cells in this ventral portion of the grafted temporal half of the eye were originally in a dorsal position in the donor eyebud. Thus, if they project in a mosaic pattern (according to their original position in the donor eyebud) they would project to the ventral tectum, and if they project according to the final regulated map, they would project to the dorsal tectum in agreement with their ventral position in the eye. By performing this procedure on animals of progressively older stages, the projection patterns of cells added to the eye at progressively later stages could be assayed. The fluorescent nerve tracing studies on the early larval N_rT_1 eyes showed that the temporoventral cells in the center of the retina projected along the lateral edge of the tectum as if they were dorsal cells. The HRP

Figure 7. HRP-labeled normal and N_rT_1 eye projections. (A) Temporodorsal fibers in a normal larval eye project to the anteroventral region of the tectum. (B) Temporoventral fibers in a normal larval eye project to the anterodorsal region of the tectum. Fibers in different regions of the normal eye have distinct projection patterns in the tectum. (C) Projection patterns of axons from cells in the temporoventral region of a N_rT_1 compound eye at stage 51. Some of the fibers are found in the anterodorsal portions of the tectum (see arrow) indicating regulation along only the dorsoventral axis. Some of the fibers still project in a mosaic fashion (anteroventrally, see asterisk) according to the original dorsal origin of the graft. (d, dorsal; v, ventral; a, anterior; p, posterior).

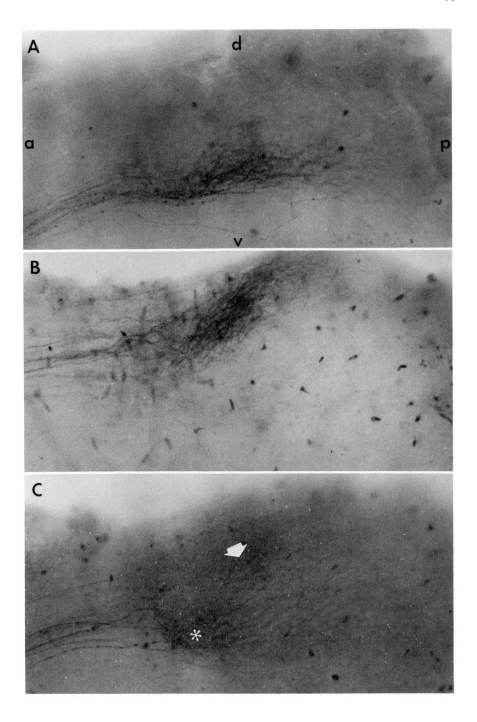

fiber tracing on temporoventral cells around the limbus of stage 49 N_rT_1 eyes showed the same mosaic projection patterns (figure 7A). In contrast, as early as stage 51-54, the temporoventral cells were found to project both to the ventral and to the dorsal tectum (figure 7C), indicating the appearance of fibers projecting in a regulated pattern. By stage 55, the peripheral retinal cells projected only in a dorsal position in the tectum as if they had become ventral cells (figure 7B). Thus, the cells in the outer part of the stage 55 N_rT_1 eye were projecting exclusively in a regulated pattern, in agreement with the pattern seen in the postmetamorphic electrophysiological maps. The regulation appears gradually in the projection patterns of the newly added cells in the outer regions of the retina.

CONCLUSION

The use of a new vital-dye fiber-tracing technique has allowed us to follow the initial stages of regulatory interactions in the **Xenopus** eyebud. Using this technique we have addressed two important issues: 1) do the cells in the embryonic eyebud possess positional cues which guide them to their proper targets in the tectum, and 2) do changes in the positional cues of early eyebud cells contribute to pattern regulation in the eyebud? Since the early fiber-tracing studies were carried out on live animals, the assayed animals could be raised to later stages, permitting the timecourse of appearance of regulated projection patterns to be documented in stages up to and including metamorphosis. This approach permitted extensions of our knowledge of pattern regulation in the eyebud beyond that obtained in previous studies in which the regulated pattern was assayed only in the postmetamorphic animals.

The in vivo fiber-tracing technique was used to investigate the presence and stability of positional cues in the developing eyebud of **Xenopus laevis**. The presence of positional cues in the early eyebud cells was tested by grafting small groups of cells into heterotopic sites in the eyebud. The results of these studies provided direct evidence that the early eyebud cells possess cell-autonomous positional cues which they retain even when completely surrounded by inappropriate neighbors. The positional cues appear to be stable under a variety of experimental conditions. In another set of experiments, the stability of the positional cues in the eyebud was challenged in compound N_rT_1 eyes, in which one half of the eye had been inverted along the dorsoventral axis. The positional cues in the grafted eyebud cells were found to be stable in all cases, even in those in which electrophysiological maps of the same eyes at later stages revealed regulated projection patterns. In both the heterotopic grafts and the compound eyes, the positional cues were stable several weeks into development. Therefore, the positional cues in the early cells appear to be indelible. Although other factors, such as neuronal activity, may also be involved, positional cues appear to play a major role in initially guiding the fibers to their proper targets in the tectum.

The results of compound N_rT_1 eye experiments have eliminated the possibility that early respecification of the positional cues in eyebud cells is involved in regulation in the eyebud. The labeled cells projected to the tectum according to their original positions in the tectum, despite the fact that regulated patterns were seen in the adult animals in 100% of the cases. The regulated projection pattern in compound eyes appeared gradually in the projection patterns of cells added to the eye at later stages. The mechanism involved in the late appearance of the regulated pattern remains a mystery. It is possible that respecification of the positional cues in the original grafted cells occurred at a later stage, and that these new positional cues were passed on to their progeny. This seems unlikely because the positional information remains stable in the central grafted cells even at stages when their progeny around the circumference of the eye are beginning to show regulated patterns. Alternatively, the positional cues in the newly added cells could be specified by signals spreading in the ciliary margin from cells of host origin. The patterns of positional cues would then be passed on to cells added to the eye subsequently, leaving the grafted cells in the center of the retina with mosaic projection patterns. This mechanism would yield a similar pattern to one in which intercalary growth at the margin of the eye gives rise to new cells with new patterns of positional values. In this case, the central grafted cells would again be expected to have mosaic projection patterns. Further knowledge of the mechanisms involved in pattern regulation will require investigations of the specific contributions of cell division and cell-cell communication to regulatory interactions. The anterograde HRP studies have revealed developmental stages when these interactions must be taking place and thus the stages which will be studied in the future.

The results of studies in which regulation in the eyebud was assayed in postmetamorphic adults have led to the proposal of several possible mechanisms which could be involved in regulatory interactions. The use of the in vivo fiber-tracing technique, which allowed us to follow the initial projection patterns of grafted eyebud cells, has allowed us to begin eliminating some of these mechanisms. Future studies concentrating on the role of specific cellular interactions in regulation in the eyebud promise to reveal the mechanisms involve in shaping and reshaping spatial patterns in the retinotectal projection.

REFERENCES

Bryant, S., V. French, and P. Bryant. 1981. Distal regeneration and symmetry. Science 212:993-1002.

Conway, K., K. Feiock, and R. K. Hunt. 1980. Polyclones and patterns in developing **Xenopus** larvae. Curr. Topics Dev. Bio. 15:216-317.

Cooke, J. and R. M. Gaze. 1983. The positional coding system in the early eye rudiment of **Xenopus laevis**, and its modification after grafting operations. J. Embryol. exp. Morph. 77:53-71.

Fernald, R. D. 1984. Vision and Behavior in an African **Cichlid** fish: Combining behavioral and physiological analyses reveals how good vision is maintained during rapid growth of the eyes. American Scientist 72:58-65.

Fraser, S. E. 1987. Intrinsic positional information guides the early formation of the retinotectal projection of **Xenopus**. Neurosci. Abst. 13(1):368.

French, V., S. Bryant, and P. Bryant. 1976. Pattern regulation in epimorphic fields. Science 158:969-981.

Fujisawa, H. 1984. Pathways of retinotectal projection in developing **Xenopus** tadpoles revealed by selective labeling of retinal axons with horseradish peroxidase. Develop. Growth and Diff. 26(6):545-553.

Fujisawa, H., N. Tani, K. Watanabe, and Y. Ibata. 1982. Branching of regenerating retinal axons and preferential selection of appropriate branches for specific neuronal connections in the newt. Dev. Bio. 90:43-57.

Fujisawa, H., K. Watanabe, N. Tani, and Y. Ibata. 1981. Retinotopic analysis of fiber pathways in the regenerating retinotectal system of the adult newt **Cynops pyroghaster**. Brain Res. 206:27-37.

Gaze, R. M. and C. Straznicky 1980. Stable programming for map orientation in disarranged embryonic eyes in **Xenopus**. J. Embryol. Exp. Morph. 55:143-165.

Gimlich, R. L. and J. Braun. 1985. Improved fluorescent compounds for tracing cell lineage. Dev. Bio. 109:509-514.

Harris, W. A. 1982. The transplantation of eyes to genetically eyeless salamanders: Visual projections and somatosensory interactions. J. Neurosci. 2:339-353.

Harris, W. A. 1984. Axonal pathfinding in the absence of normal pathways and impulse activity. J. Neurosci. 4:1153-1162.

Hollyfield, J. G. 1971. Differential growth of the neural retina in **Xenopus laevis** larvae. Dev. Bio. 24:264-286.

Holt, C. E. 1980. Cell movements in **Xenopus** eye development. Nature 28:850-852.

Holt, C. E. 1984. Does timing of axon outgrowth influence initial retinotectal topography in **Xenopus**? J. Neurosci. 4:1130-1152.

Holt, C. E. and W. A. Harris. 1983. Order in the initial retinotectal map in **Xenopus**: A new technique for labeling growing nerve fibres. Nature 301:150-152.

Hunt, R. K. and M. Jacobson. 1972. Development and stability of positional information in **Xenopus** retinal ganglion cells. Proc. Nat. Acad. Sci. USA 69:780-783.

Hunt, R. and M. Jacobson. 1973. Neuronal locus specificity: Altered pattern of spatial deployment in fused fragments of embryonic **Xenopus** eyes. Science 180:509-511.

Ide, C. F., P. Reynolds, and R. Tompkins. 1984. Two healing patterns correlate with different neural connectivity patterns in regenerating embryonic **Xenopus** retina. J. Exp. Zool. 230:71-80.

Ide, C. F., L. Wunsh, P. Lecat, D. Kahn, and E. Noelke. 1987. Healing modes correlate with visuotectal pattern formation in regenerating embryonic **Xenopus** retina. Dev. Bio. 124:316-330.

Jacobson, M. 1968. Development of neuronal specificity in retinal ganglion cells of **Xenopus**. Dev. Bio. 17:202-218.

Jacobson, M. 1976. Histogenesis of retina in the clawed frog with implications for the pattern of development of retinotectal connections. Dev. Bio. 103:541-545.

Meyer, R. L. 1984. Target selection by surgically misdirected optic fibers in the tectum of goldfish. J. Neurosci. 4:234-250.

Nieuwkoop, P. D. and J. Faber. 1956. Normal Table of **Xenopus laevis**. (Daudin) Elsevier-North Holland Publishing Co., Amsterdam.

O'Gorman, S., J. Kilty, and R. K. Hunt. 1987. Healing and growth of half eye "compound eyes" in **Xenopus**: Application of an interspecific cell marker. J. Neurosci. 7(11):3764-3782.

O'Rourke, N. A. and S. E. Fraser. 1986a. Dynamic aspects of retinotectal map formation as revealed by a vital-dye fiber-tracing technique. Dev. Bio. 114:265-276.

O'Rourke, N. A. and S. E. Fraser. 1986b. Pattern regulation in the eyebud of **Xenopus** studied with a vital-dye fiber tracing technique. Dev. Bio. 114:277-288.

O'Rourke, N. A. and S. E. Fraser. 1986c. Gradual appearance of a regulated projection pattern in the developing eyebud of **Xenopus laevis**. Neurosci. Abst. 12:543.

Sakaguchi, D. S. and R. K. Murphey. 1985. Map formation in the developing **Xenopus** retinotectal system: an examination of ganglion cell terminal arborizations. J. Neurosci. 5:3228-3245.

Straznicky, C. and R. M. Gaze. 1971. The growth of the retina in **Xenopus laevis**: An autoradiographic study. J. Embryol. Exp. Morph. 26:67-79.

Thiebaud, C. H. 1983. A reliable new cell marker in **Xenopus**. Dev. Bio. 98:245-249.

Wolpert, L. 1969. Positional information and the spatial pattern of cellular differentiation. J. Theor. Bio. 25:1-47.

Wolpert, L.   1971.   Positional information and pattern formation.
    Curr. Top. Dev. Bio.   6:183-224.
Yoon, M. G.   1975.   Topographic polarity of the optic tectum studied
    by reimplantation of the tectal tissue in adult goldfish.   Cold
    Spring Harbor Symp. Quant. Bio. 40:503-519.

Factors Underlying Loss of Retinal Ganglion Cells
Leo M. Chalupa

Within the last decade or so, substantial information has accumulated on the prenatal development of the retina and the visual pathways (for recent review, see Chalupa & White, 1988). In this chapter I would like to focus upon one of the most dramatic events that occurs during the formation of the mammalian visual system, the loss of retinal ganglion cells. In the cat, the species my laboratory has been studying, approximately 5 of 6 ganglion cells generated during fetal life do not survive to maturity!

What could cause such massive elimination of developing neurons? Cell death appears to be a ubiquitous phenomenon during the ontogenesis of the mammalian nervous system, and the factors underlying this regressive event have been considered from various perspectives (for reviews see: Clarke, 1985; Oppenheim, 1985; Williams & Herrup, 1988). A major emphasis has been given in this literature to the role of the target in the regulation of neuronal death: cells which die are commonly thought to have lost a competitive interaction for some essential target-derived entity. This explanation could also account for the elimination of early projection errors since aberrant neurons projecting to the inappropriate target would be ill suited to compete effectively with the presumably more robust normal connections. The emphasis on axon-target interactions during early development is in keeping with the prevalent notion that specificity in a complex nervous system cannot be preprogrammed entirely by the genome. Rather, experiental factors, including the spontaneous activity of cells before the birth of the organism, mold genetically programmed events to form the precise connections evident in the mature nervous system (e.g., Changeux & Danchin, 1976).

The foregoing explanation of cell loss undoubtedly contains a fair degree of validity. Nevertheless, the evidence that has accumulated on the early development of the mammalian visual system suggests quite clearly that axon-target interactions alone cannot explain ontogenetic loss of retinal ganglion cells. There are two major reasons why this is the case. First, the viewpoint summarized above implies that developmental cell loss reflects mainly a corrective process whereby various types of early projection errors are eliminated. Unquestionably, the connections of the developing visual system are less precise than those present at maturity. However, as will be discussed below, in the case of the mammalian visual system, there is now reason to believe that we have been overly exuberant about the degree of exuberance (or imprecision) manifested by immature projections. Second, the foregoing account unduly emphasizes competitive interactions among outgrowing axons at targets as the primary mechanism for the loss of cells. As will be discussed, there is reason to believe that interactions among dendrites also play an important role in retinal development, and I will propose that such interactions are an integral factor in the control of ganglion cell loss.

In the following sections, I will first briefly review the evidence documenting ganglion cell loss during the early development of the mammalian retina. I will then consider the developmental events that might be related to the loss of these neurons. Finally, I will propose a model to explain the elimination of retinal ganglion cells.

EVIDENCE FOR LOSS OF RETINAL GANGLION CELLS

Three lines of evidence have demonstrated retinal ganglion cell loss during early development of the visual system. First, numerous pyknotic profiles have been observed in the ganglion cell layer of the developing retina (e.g., Cunningham et al., 1982; Sengelaub & Finlay, 1982). Second, counts of ganglion cells or of presumed ganglion cell precursors (e.g., Stone et al., 1982; Perry et al., 1983) have been reported to be substantially higher early in development than at maturity. Third, estimates of the total population of fibers in the developing optic nerve have revealed an initial overproduction followed by an elimination of these axons (e.g., Rakic & Riley, 1983a; Williams et al., 1983, 1986). Each of these approaches for estimating the time course and the magnitude of ganglion cell loss has some limitations as well as advantages (cf., Chalupa & White, 1988). Nevertheless, the available evidence permits several generalizations regarding retinal ganglion cell loss and these will be summarized below.

An overproduction and loss of retinal ganglion cells has been noted in diverse species including: the chick (Rager & Rager, 1976), rat (Cunningham et al., 1982), opposum (Kirby & Wilson, 1984), cat (Ng & Stone, 1982; Williams et al., 1983, 1986), rhesus monkey (Rakic & Riley, 1983a) and human (Provis et al., 1985). In contrast, in amphibians and in fish, retinal ganglion cells are generated throughout life without significant loss of these neurons (Easter & Stuermer, 1984; Coleman et al., 1984). It is quite likely, therefore, that loss of retinal ganglion cells occurs in all vertebrates in which the generation of retinal ganglion cells ceases at a particular developmental period.

The general time course of retinal ganglion cell loss appears quite comparable among different species. There are two major components to this phenomenon: a relatively brief period during which many cells are lost followed by a more prolonged period during which fewer cells are eliminated. This is illustrated in Figure 1, which depicts the addition and attrition of axons in the optic nerve of the cat. It should be noted that the peak number of fibers found in the developing optic nerve (at about embryonic day 39) underestimates the total number of axons produced during development because some axons have already been eliminated before the peak count is attained. This is indicated by the presence of necrotic fibers within the fetal optic nerve at a time when growth cones are also present and when the number of axons is still increasing. The first necrotic fibers are observed in the cat's optic nerve as early as E28, some 11 days before the peak number of fibers is attained. Based on a 1-hour estimate for the clearance time of single axon's debris, Williams et al. (1986) concluded that between 100,000 and 200,000 axons are lost before the peak population of approximately 700,000 is attained. It is reasonable to assume that during development, as in adulthood (e.g., Chalupa et al., 1984), estimates of the fiber population in the cat's optic nerve correspond quite closely to estimates of ganglion cell number (cf., Williams et al., 1986, Lia et al., 1986). This means that a total of some 900,000 retinal ganglion cells are generated in the fetal cat retina, of

72

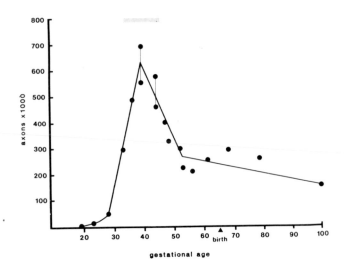

Figure 1. The overproduction and elimination of fibers in the optic nerve of the cat. Each data point represents one animal. As indicated, gestation in the cat is about 65 days. The gray bar denotes the estimated total number of axons produced during fetal development. The figure is reproduced from Williams et al. (1986).

which only 150,000 to 160,000 survive to adulthood.

In primates (Rakic & Riley, 1983a; Provis et al., 1985), the developmental overproduction of these cells, as estimated from axon counts, is two- or three-fold. The same magnitude of overproduction appears to be the case in rodents (Perry et al., 1983), whereas in the chicken (Rager & Rager, 1976) it is considerably less than two-fold. To some degree the species differences may reflect procedural variations among experimenters. For instance, with the exception of Williams et al. (1986), estimates of the the total number of fibers produced during development have not taken into account the axons eliminated prior to the attainment of the peak population. Additionally, the accuracy of the sampling methods used to obtain fiber population estimates also probably varies across studies. In spite of these reservations, it is clear that there is a massive loss of ganglion cells during the early development of the mammalian visual system.

Probably all ganglion cell loss is the result of cell death rather than the transformation of immature ganglion cells into some other cell type. The latter idea was put forth by Hinds and Hinds (1978, 1983), who hypothesized that some ganglion cell precursors withdraw their axons from the developing optic nerve and become transformed into displaced amacrine cells. This notion is quite attractive because it would explain in a parsimonious manner the presence of a substantial population of displaced amacrine cells in the ganglion cell layer. Recent studies (Wässle et al., 1987; Wong & Hughes, 1987a) have reported that the number of displaced amacrine cells in the adult cat retina is about five times greater than the number of ganglion cells.

There have been several attempts to obtain evidence that displaced amacrine cells have transient axons that were subsequently retracted from the optic nerve (Perry et al., 1983; Dreher et al., 1983; Campbell et al., 1987). In each case the results did not support the transformation hypothesis. However, this matter is not entirely resolved because all studies have assumed that axons of amacrine cell precursors would grow a considerable distance along the optic pathway, past the optic chiasm and even into the main retinorecipient targets. Such an assumption was not made by Hinds and Hinds. If the axons of displaced amacrine cell precursors only navigated a short distance into the optic nerve, it would be very difficult, given the methodology available currently, to demonstrate unequivocally their presence. However, until there is evidence to the contrary, it seems reasonable to assume that ganglion cell loss is due to cell death.

DEVELOPMENTAL EVENTS RELATED TO GANGLION CELL LOSS

In this section I will consider the degree to which the formation of certain fundamental features of the visual system can be related to the loss of retinal ganglion cells. Specifically, the following features will be considered: (a) the segregation of projections from the two eyes within the main retinorecipient nuclei, (b) the establishment of the partial retinal decussation pattern, (c) the topographic organization of early retinal projections, and (d) the establishment of regional density gradients in the distribution of ganglion cells across the surface of the retina.

SEGREGATION OF RETINAL PROJECTIONS.

It is now well established that projections from the two eyes are initially intermingled within the lateral geniculate nucleus and the superior colliculus before segregating to form distinct ocular domains (e.g., Rakic, 1976; Williams & Chalupa, 1982; Shatz, 1983). In the fetal cat some 200,000 axons are eliminated from each optic nerve during the segregation process (Williams et al., 1986). This suggests that retinal ganglion cell death could contribute to the segregation of retinal inputs: ganglion cells would die if they initally projected to the inappropriate ocular domain within the lateral geniculate or the superior colliculus. However, a massive number of ganglion cells is lost before there is any hint of ocular segregation in these retinorecipient nuclei, and some cell loss continues after segregation has been completed.

Direct evidence that the segregation process is at least partially due to retinal ganglion cell loss is provided by the results of prenatal enucleation experiments. Removal of one eye, at a time when projections from the two eyes are intermingled, results in an apparent maintenance of the widespread projection pattern from the remaining eye (Rakic, 1981; Land & Lund, 1979; Chalupa & Williams, 1984; White et al., 1987). In terms of the issue under consideration, the relevant point is how this in utero manipulation affects the size of the ganglion cell population in the remaining eye. In the cat (Williams et al., 1983; Chalupa et al., 1984) and monkey (Rakic & Riley, 1983b), it has been demonstrated that prenatally enucleated animals have a greater than normal number of fibers in the remaining optic nerve. However, the magnitude of this increase is modest, about 20% in cat and 25% in the monkey. In absolute numbers, prenatal eye removal in the fetal cat saves about 30,000 cells from elimination.

The foregoing observations indicate that ganglion cell loss plays a role in the segregation of initially intermingled retinal projections, although other factors such as the restructuring of immature retinogeniculate terminal arbors (Sretavan & Shatz, 1986) contribute to this process. The evidence also clearly indicates that the vast majority of retinal ganglion cells are lost for some other reasons.

ESTABLISHMENT OF RETINAL DECUSSATION.

To examine the formation of partial retinal decussation patterns, we have made injections of various tracers into the retinorecipient nuclei or the optic tract of one hemisphere in fetal cats (Lia et al., 1983, 1987a) and in fetal rhesus monkeys (Chalupa & Lia, in preparation). These studies have revealed that the partial decussation patterns characteristic of mature animals are apparent very early in fetal development, shortly after retinal fibers have innervated their target nuclei and before extensive loss of ganglion cells has occurred. In fetal monkeys, and even more so in fetal cats, some cells do project to the inappropriate hemisphere. However, the number of such neurons in each hemi-retina is never very large so that only a minor degree of ganglion cell loss can be related to corrections of immature decussation patterns.

RETINOTOPIC ORDER OF EARLY PROJECTIONS.

In another series of experiments we have examined the retinotopic order of prenatal retinal projections (Ostrach et al., 1986). This was done by making focal deposits of retrograde tracers into the superior colliculus of fetal cats at known gestational ages and subsequently charting the loci of labeled cells in the ipsilateral and the contralateral retina. At all fetal ages, a high density of labeled cells was found to be confined to a limited region of each retina. Further, the locus of this region shifted in an orderly manner with the placement of the tracer within different region of the colliculus, reflecting a retinotopic order characteristic of the mature animal. In all cases there were some labeled cells scattered throughout each retina, but the number of such ectopic neurons was always small, less than 1% of the total number of cells labeled in the high density region. While developmental refinements in the retinotopic organization of the visual system do occur, our observations indicate that the projections of the immature visual system exhibit a remarkable degree of order. Consequently, only a small percentage of the retinal ganglion cell loss that normally occurs during early development could reflect a correction of topographic errors.

RETINAL REGIONAL SPECIALIZATION.

In the adult cat, as in other animals with well developed focal vision, the distribution of ganglion cells across the retinal surface is non-uniform. The peak density of these neurons in the area centralis is about 80 times greater than the density in the periphery of the retina. Several years ago, in a collaborative effort with Jonathan Stone and David Rapaport, Rob Williams and I (Stone et al., 1982) discovered that in the fetal cat retina the distribution of ganglion cells was relatively uniform. One of the explanations we put forth for the attainment of the mature retinal distribution pattern was cell death: cells in the periphery of the retina were eliminated in substantially greater numbers than cells in the central region. There have been several attempts recently to test the validity of this hypothesis in a variety of mammalian species (Dunlop et al., 1987; McCall et al., 1987; Provis, 1987; Robinson, 1987).

A recent study from my laboratory (Lia et al., 1987b) has demonstrated that the primary mechanism for the prenatal formation of the central to peripheral variation in the density of ganglion cells is not cell death, but rather the non-uniform expansion of the growing prenatal retina. During fetal life the peripheral regions of the retina expand much more than the central region, and this effectively dilutes the density of cells in the periphery more than in the center. The non-uniform growth of the retina continues after birth, as originally documented by Mastronade et al. (1984), further increasing the cell density gradient.

It should be stressed that our findings do not exclude the possiblity that cell death may be involved in the refinement of regional variations in the distribution of ganglion cells across the retinal surface (cf., Wong & Hughes, 1987b). However, cell death does not play a major role in establishing the central-to-peripheral gradient in ganglion cell density. This is now firmly established in the cat, and I suspect that non-uniform retinal growth will be shown to be the main factor in forming retinal regional gradients in all mammals. In ongoing work we are now in the process of examining this problem in the fetal rhesus monkey (Lia & Chalupa, 1988).

DENDRITIC INTERACTIONS

Prior to discussing a model to account for the bulk of retinal ganglion cell loss, it is necessary to consider one additional characteristic of developing ganglion cells: interactions among developing dendritic processes. The concept of dendritic interactions stems from the elegant studies of Heinz Wässle and his colleagues (Wässle & Reimann, 1978; Wässle et al., 1981a,b), who showed that ganglion cells in the adult retina are distributed in mosaic patterns. This means that a given ganglion cell class provides complete coverage of the retinal surface with only limited overlap among dendritic arbors. As pointed out by Wässle and Reimann (1978), such an organization at maturity suggests that during development there must have been interactions among dendritic processes of a given ganglion cell class.

There are two different ways of viewing such dendritic interactions. One possibility is that growing dendrites recognize the membranes of "conspecific" cells and such recognition impedes further dendritic overlap. The other possibility is that dendritic interactions result from competition among maturing dendrites for afferent inputs or for trophic factors that may be provided by the afferent cells (Perry & Linden, 1982; Linden & Perry, 1982).

While the mechanisms underlying dendritic interactions are still unknown, there is now considerable evidence that such interactions do occur during retinal development. For instance, a recent study from my laboratory (Kirby & Chalupa, 1986) showed that the dendritic fields of alpha cells are smaller than normal in the remaining eye of adult cats that were monocularly enucleated before birth. This finding supports the Wässle hypothesis because the density of ganglion cells in the retina of the enucleated animals is greater than normal, and such "crowding" would lead to smaller dendritic fields if there were dendritic interactions during early development. Others have demonstrated that manipulations which reduce the density of ganglion cells result in dendritic fields that are greater than normal (Linden & Perry, 1982; Ault et al., 1985).

A MODEL TO EXPLAIN RETINAL GANGLION CELL LOSS

The model I propose to account for the bulk of ganglion cell loss is illustrated schematically in figure 2. In the top panel is a group of retinal ganglion cells of the same subclass (e.g., on-center alpha cells) with ingrowing axons directed towards their target nucleus. The somas of these "conspecific" neurons are clustered in the same retinal region. As depicted in the second panel, one of the axons arrives in the vicinity of the target before the axons of the other cells. The innervation of the target is hypothesized to provide a trophic substance for this cell which is incorporated by the terminal and transported to the perikaryon. This trophic factor nourishes the cell, so that when dendritic growth commences the dendritic tree of this particular neuron will grow much more vigorously than those of the neighboring cells whose axons arrive later at the target or are still in the process of navigating towards the target. Due to inhibitory interactions among the conspecific neurons in this retinal region, such growth will retard the dendritic extensions of surrounding cells. The model further assumes that in order for a ganglion cell to survive, a certain minimal dendritic field dimension must be attained at a critical point in development. For this reason, as development proceeds, the conspecific cells surrounding the neuron with the flourishing dendritic tree will perish (lower panel of the figure).

Dendritic field size of all ganglion cell classes varies as a function of retinal eccentricity (Boycott & Wässle, 1974). This reflects variations in cell density, although there is reason to believe that factors intrinsic to ganglion cells also govern the dimensions of dendritic fields (Eysel et al., 1985; Kirby & Chalupa, 1986; Schall & Leventhal, 1987; Dann et al., 1987). The fact that dendritic fields vary in size regionally in the mature retina would seem to imply that the critical field size required to maintain the survival of these cells differs across the retina. However, this is not the case, as most retinal ganglion cells are eliminated before there is substantial non-uniform growth of the fetal retina (see Lia et al., 1987b). Thus, ganglion cell death occurs predominantly at a time when the distribution of these neurons across the retinal surface is relatively uniform. This means that the critical dendritic field dimension required for survival would be relatively constant for all retinal regions. After this period of ganglion cell elimination the regional density gradient of ganglion cells and the concomitant changes in the size of dendritic fields result primarily from the rapid non-uniform growth of the retina. Since the peripheral retina stretches much more than the central region,

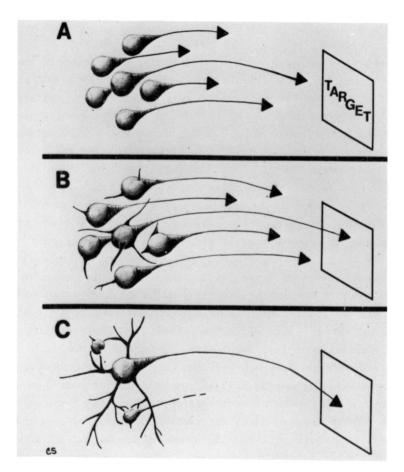

Figure 2. A schematic representation of the model accounting for the loss of ganglion cells in the developing retina. See text for explanation.

any uneveness in the elasticity of the peripheral retina (which might be expected from such factors as the vascularization pattern) would lead to a greater irregularity in the mosaic pattern of peripheral ganglion cells (cf., Wässle & Reimann, 1978).

I will now briefly consider the four main assumptions of the model:

1. There are significant differences among cells of a given class in a particular retinal region in the arrival time of their axons at the target nucleus. The relevant information about the ingrowth of specific classes of ganglion cell axons is lacking, but this assumption appears plausible given our current knowledge of guidance mechanisms and process extension in the developing nervous system. In particular, _in vivo_ studies of retinal ganglion cell

axon growth cones have revealed substantial variations in the rate of axon outgrowth as well as in the directness of the process extension towards the appropriate target cells (e.g., Williams et al., 1986; Bovolenta & Mason, 1987). Slight timing differences in completing the mitotic cycle, among neurons of a given class at a given retinal region, could also contribute to the variability of axonal outgrowth.

2. Innervation of the target provides a trophic factor which enhances growth of the perikaryon in the target cell. In contemporary cellular neurobiology, the notion that trophic substances, released by target cells, play a critical role in the maintenance of ingrowing cells is well entrenched. While at present there is no direct evidence for a trophic substance being involved in the development of the mammalian visual system, this has been inferred from studies which have demonstrated that target cells are essential for the survival of developing retinal ganglion cells (e.g., Carpenter et al., 1986). The assumption I have made builds on this inference: not only is the target cell essential for the maintenance of developing retinal projections, it also provides a substance which enhances dendritic growth of retinal ganglion cells.

3. There are inhibitory interactions among developing ganglion cells of a specific class. The presence of retinal mosaics, the fact that dendritic field dimensions in the normal retina vary as a function of ganglion cell density, and the modifiability of dendritic field size by developmental manipulations affecting cell density all lend support to this assumption. As noted previously, it is unclear if such interactions are due to some type of recognition among dendrites of a specific cell class or whether it involves a competition for afferents. Synaptic input from bipolar and amacrine cells is not a factor in the initial outgrowth and elaboration of ganglion cell dendrites as many ganglion cells have prominent dendritic fields prior to the appearance of synapses in the inner plexiform layer (Lia et al., 1983; Maslim et al., 1986; Ramoa et al., 1987). However, the contribution of retinal afferents to the early growth of ganglion cell dendritic processes cannot be excluded since this might be mediated by trophic agents provided by presynaptic neurons. There is even evidence that such agents could be provided by retinal glial cells (e.g., Raju & Bennett, 1986). Furthermore, when inner plexiform layer synaptic contacts do become established, this could contribute to the morphological

restructuring of ganglion cell dendrites (Maslim et al., 1986; Romoa et al., 1987; Dann et al., 1987). It would be of considerable interest to study the contribution of retinal afferents to ganglion cell dendritic morphology by examining dendritic morphology in a mature retina which lacked bipolar and/or amacrine neurons. Since ganglion cells are born prior to the other retinal neurons (e.g., Polley et al., 1981, 1985), this might be achieved by the use of anti-mitotic agents or by focal X-radiation methods.

4. To survive to maturity ganglion cells of a given class must attain a minimal dendritic field size. This is the most speculative of the assumptions underlying the model. Why should the size of the dendritic tree determine the survival of a retinal ganglion cell? One possibility is that trophic factors from retinal afferents presynaptic to the ganglion cells are required for long-term survival. The uptake of such factors by cells with subnormal dendritic trees would be insufficient to maintain these neurons. Such cells may also have lower levels of electrical activity than their neighbors and this could be a contributory factor in their demise (cf., Lipton, 1986).

Several observations on the developing visual system are consistent with the model I have described. First, the temporal sequence of certain key developmental events is in keeping with the model. For instance, in the fetal cat the initial innervation of the lateral geniculate nucleus and the superior colliculus by retinal ganglion cell axons (Shatz, 1983; Williams & Chalupa, 1982) takes place before the outgrowth of dendritic processes in these cells. Furthermore, the acute phase of ganglion cell loss (Williams et al., 1986) occurs at a time when dendritic processes of many ganglion cells are flourishing (Maslim et al., 1986). Second, the model accounts for degenerating growth cones within the fetal optic nerve (see figure 19 of Williams et al. 1986). They would be expected if some ganglion cells died, due to insufficient dendritic growth, even while their axons were still in the process of navigating toward the targets. While only a few such necrotic growth cones were observed in the optic nerve, I suspect that many cells are eliminated before their processes reach the vicinity of the target. This might explain why injections of retrograde tracers into retinorecipient targets and even into the optic chiasm of the developing brain label fewer retinal ganglion cells than the number of fibers in the midorbital portion of the optic nerve (Perry et al., 1983; Lia et al., 1987b).

Third, the model also explains why monocular enucleation in fetal life results in only a modest increase in the number of ganglion cells in the remaining eye. Since prenatal enucleation presumably doubles the number of target cells available to the axons of the remaining eye, one might expect that this would result in a doubling of the ganglion cell population in the remaining retina. However, the number of cells saved is reasonably stable, in spite of wide developmental fluctuations in the number of cells that are present at the time the enucleation is performed. For instance, eye removal at E40, when there are more than 600,000 retinal ganglion cells, results in about the same degree of savings as does enucleation 10 days later, when the ganglion cell population has dropped to about 300,000 cells. In all these cases the size of the remaining retina at maturity is no greater than normal (Chalupa et al, 1984). According to the model, retinal size would limit the number of surviving cells due to the requirement that they attain a minimal dendritic field size at a certain stage of development. The fact that early eye removal saves some ganglion cells from elimination demonstrates that afferent-target competititon also plays a role in cell survival, but according to the model, cell survival reflects a balance of this factor with the dendritic interactions occurring in the retina.

Finally, I wish to stress that the model is not intended to explain all retinal ganglion cell loss. The available evidence clearly indicates that such cell loss results from multiple factors, including the events discussed in this chapter, namely: the segregation of retinal afferents at the target nuclei, the correction of early topographic and decussation errors, and the refinement of retinal regional specialization. No doubt, the concepts put forth in the model will need to be modified as our knowledge of the events occurring during early development of the mammalian visual system continues to expand. I believe, however, that the ultimate explanation of ganglion cell survival will include both axon-target as well as retinal dendritic interactions.

ACKNOWLEDGEMENTS

I thank Cheryl White and Barry Lia for their comments on earlier drafts of this chapter. I also thank Cara Snider for making the illustrations shown in figure 2. The work in my laboratory on the prenatal development of the visual system is supported by NIH/NEI grant EY03391 and NSF grant INT-8320440. This chapter was written while I was a Fellow-in-Residence at the Neurosciences Institute.

REFERENCES

Ault, S.J., J.D. Schall and A.G. Leventhal. 1985. Experimental alteration of cat retinal ganglion cell dendritic field structure. Soc. Neurosci. Abtsr. 11:15.

Bovolenta, P. and C. Mason. 1987. Growth cone morphology varies with position in the developing mouse visual pathway from retina to first targets. J. Neurosci. 7(5):1447-1460.

Boycott, B.B. and H. Wässle. 1974. The morphological types of ganglion cells of the domestic cat's retina. J. Physiol. 240:397-419.

Campbell, G., A.S. Ramoa and C.J. Shatz. 1987. Do amacrine cells extend, then retract a centrally-projecting axon? Soc. Neurosci. Abstr. 13:589.

Carpenter, P., A.J. Sefton, B. Dreher and W.-L. Lim. 1986. Role of target tissue in regulating the development of retinal ganglion cells in the albino rat: Effects of kainate lesions in the superior colliculus. J. Comp. Neurol. 251:240-259.

Chalupa, L.M. and R.W. Williams. 1984. Organization of the cat's lateral geniculate nucleus following interruption of prenatal binocular competition. Human Neurobiol. 3:103-107.

Chalupa, L.M. and C.A. White. 1988. Prenatal development of the mammalian visual system. In: Development of Sensory Systems in Mammals, J.R. Coleman, ed., John Wiley & Sons, Inc., in press.

Chalupa, L.M., R.W. Williams and Z. Henderson. 1984. Binocular interaction in the fetal cat regulates the size of the ganglion cell population. Neurosci. 12:1139-1146.

Changeux, J.-P. and A. Danchin. 1976. Selective stabilisation of developing synapses as a mechanism for the specification of neuronal networks. Nature. 264:705-712.

Clarke, P.G.H. 1985. Neuronal death in the development of the vertebrate nervous system. Trends in Neurosci. 8:345-349.

Coleman, L-A., S.A. Dunlop and L.D. Beazley. 1984. Patterns of cell division during visual streak formation in the frog Limnodynastes dorsalis. J. Embryol. exp. Morph. 83:119-135.

Cunningham, T.J., I.M. Mohler and D.L. Giordano. 1982. Naturally occurring neuron death in the ganglion cell layer of the neonatal rat: Morphology and evidence for regional correspondence with neuron death in superior colliculus. Dev. Brain Res. 2:203-215.

Dann, J.F., E.H. Buhl and L. Peichl. 1987. Dendritic maturation in cat retinal ganglion cells: a Lucifer yellow study. Neurosci. Letters 80:21-26.

Dreher, B., R.A. Potts and M.R. Bennett. 1983. Evidence that the early postnatal reduction in the number of rat retinal ganglion cells is due to wave of ganglion cell death. Neurosci. Letters 36:255-260.

Dunlop, S.A., W.A. Longley and L.D. Beazley. 1987. Development of the area centralis and visual streak in the grey kangaroo Macropus fuliginosus. Vision Res. 27:151-164.

84

Easter, S.S., Jr. and C.A.O. Struermer. 1984. An evaluation of the hypothesis of shifting terminals in goldfish optic tectum. J. Neurosci. 4:1052-1063.

Eysel, U.T., L. Peichl and H. Wässle. 1985. Dendritic plasticity in the early postnatal feline retina: Quantitative characteristics and sensitive period. J. Comp. Neurol. 242:134-145.

Hinds, J.W. and P.L. Hinds. 1978. Early development of amacrine cells in the mouse retina: An electron microscopic, serial section analysis. J. Comp. Neurol. 179:277-300.

Hinds, J. W. and P.L. Hinds. 1983. Development of retinal amacrine cells in the mouse embryo: Evidence for two modes of formation. J. Comp. Neurol. 213:1-23.

Kirby, M.A. and P.D. Wilson. 1984. Axon count in the developing optic nerve of the North American opossum: Overproduction and elimination. Soc. Neurosci. Abts. 10:467.

Kirby, M.A. and L.M. Chalupa. 1986. Retinal crowding alters the morphology of alpha ganglion cells. J. Comp. Neurol. 251:532-541.

Land, P.W. and R.D. Lund. 1979. Development of the rat's uncrossed retinotectal pathway and its relation to plasticity studies. Science 205:698-700.

Lia, B., R.W. Williams and L.M. Chalupa. 1983. Early development of retinal specialization: The distribution and decussation patterns of ganglion cells in the prenatal cat demonstrated by retrograde peroxidase labeling. Soc. Neurosci. Abstr. 9:702.

Lia, B., R.W. Williams and L.M. Chalupa. 1986. Does axonal branching contribute to the overproduction of optic nerve fibers during early development of the cat's visual system? Dev. Brain Res. 25:296-301.

Lia, B., M.A. Kirby and L.M. Chalupa. 1987a. Decussation of retinal ganglion cell projections during prenatal development of the cat. Soc. Neurosci. Abstr. 13:1690.

Lia, B., R.W. Williams and L.M. Chalupa. 1987b. Formation of retinal ganglion cell topography during prenatal development. Science 236:848-851.

Lia, B. and L.M. Chalupa. 1988. Prenatal development of regional specializaion in the primate retina. Invest. Ophthal. Vis. Sci. suppl. 29:378.

Linden, R. and V.H. Perry. 1982. Ganglion cell death within the developing retina: A regulatory role for retinal dendrites? Neuroscience 7:2813-2827.

Lipton, S.A. 1986. Blockade of electrical activity promotes the death of mammalian retinal ganglion cells in culture. Proc. Natl. Acad. Sci. USA. 83:9774-9778.

Maslim, J., M. Webster and J. Stone. 1986. Stages in the structural differentiation of retinal ganglion cells. J. Comp. Neurol. 254:382-402.

Mastronarde, D.N., M.A. Thibeault and M.W. Dubin. 1984. Non-uniform postnatal growth of the cat retina. J. Comp. Neurol. 228:598-608.

McCall, M. J., S.R. Robinson and B. Dreher. 1987. Differential retinal growth appears to be the primary factor producing ganglion cell density gradient in the rat. Neurosci. Letters 79:78-84.

Ng, A.Y.K. and J. Stone. 1982. The optic nerve of the cat: Appearance and loss of axons during normal development. Dev. Brain Res. 5:263-271.

Oppenheim, R.W. 1985. Naturally occurring cell death during neural development. Trends in Neurosci. 8:487-493.

Ostrach, L.H., M.A. Kirby and L.M. Chalupa. 1986. Topographic organization of retinocollicular projections in the fetal cat. Soc. Neurosci. Abstr. 12:119.

Perry, V.H. and R. Linden. 1982. Evidence for dendritic competition in the developing retina. Nature 297:683-685.

Perry, V.H., Z. Henderson and R. Linden. 1983. Postnatal changes in retinal ganglion cell and optic axon populations in the pigmented rat. J. Comp. Neurol. 219:356-368.

Polley, E.H., C. Walsh and T.L. Hickey. 1981. Neurogenesis in cat retina. A study using ^{3}H-thymidine autoradiography. Soc. Neurosci. Abstr. 7:672.

Polley, E.H., R.P. Zimmerman and R.L. Fortney. 1985. Development of the outer plexiform layer (OPL) of the cat retina. Soc. Neurosci. Abstr. 11:14.

Provis, J.M., D. Van Driel, F.A. Billson and P. Russell. 1985. Human fetal optic nerve: Overproduction and elimination of retinal axons during development. J. Comp. Neurol. 233:429-451.

Provis, J. M. 1987. Patterns of cell death in the ganglion cell layer of the human fetal retina. J. Comp. Neurol. 259:237-246.

Rager, G. and U. Rager. 1976. Generation and degeneration of retinal ganglion cells in the chicken. Exp. Brain Res. 25:551-553.

Raju, T.R. and M. R. Bennett. 1986. Retinal ganglion cell survival requirements: A major but transient dependence on Müller glia during development. Brain Res. 383:165-176.

Rakic, P. 1976. Prenatal genesis of connections subserving ocular dominance in the rhesus monkey. Nature 261:467-471.

Rakic, P. 1981. Development of visual centers in the primate brain depends on binocular competition before birth. Science 214:928-931.

Rakic, P. and K.P. Riley. 1983a. Overproduction and elimination of retinal axons in the fetal rhesus monkey. Science 219:1441-1444.

Rakic, P. and K.P. Riley. 1983b. Regulation of axon number in primate optic nerve by prenatal binocular competition. Nature 305:135-137.

Ramoa, A.S., G. Campbell and C.J. Shatz. 1987. Transient morphological features of identified ganglion cells in living fetal and neonatal retina. Science 237:522-525.

Robinson, S. R. 1987. Ontogeny of area centralis in the cat. J. Comp. Neurol. 255:50-67.

Schall, J.D. and A.G. Leventhal. 1987. Relationship between ganglion cell dendritic structure and retinal topography in the cat. J. Comp. Neurol. 257:149-159.

Sengelaub, D.R. and B.L. Finlay. 1982. Cell death in the mammalian visual system during development: I. Retinal ganglion cells. J. Comp. Neurol. 204:311-317.

Shatz, C.J. 1983. The prenatal development of the cat's retinogeniculate pathway. J. Neurosci. 3:482-499.

Sretavan, D.W. and C.J. Shatz. 1986. Prenatal development of retinal ganglion cell axons: Segregation into eye-specific layers within the cat's lateral geniculate nucleus. J. Neurosci. 6:234-251.

Stone, J., D.H. Rapaport, R.W. Williams and L.M. Chalupa. 1982. Uniformity of cell distribution in the ganglion cell layer of prenatal cat retina: Implications for mechanisms of retinal development. Dev. Brain Res. 2:231-242.

Wässle, H. and H.J. Reimann. 1978. The mosaic of nerve cells in mammalian retina. Proc. Roy. Soc. Lond. B. 200:441-461.

Wässle, H. L. Peichl and B.B. Boycott. 1981a. Morphology and mosaic of the on- and off-beta cells in the cat retina and some functional considerations. Proc. Roy. Soc. Lond. B. 212:177-195.

Wässle, H., L. Peichl and B.B. Boycott. 1981b. Dendritic territories of cat retinal ganglion cells. Nature 292:344-345.

Wässle, H., M. H. Chun and F. Müller. 1987. Amacrine cells in the ganglion cell layer of the cat retina. J. Comp. Neurol. 265:391-408.

White, C., L.M. Chalupa, M.A. Kirby, B. Lia and L. Maffei. 1987. Functional consequences of interrupting prenatal binocular interactions in the dorsal lateral geniculate nucleus of the cat. Soc. Neurosci. Abstr. 13:1536.

Williams, R.W. and L.M. Chalupa. 1982. Prenatal development of retinocollicular projections in the cat: An anterograde tracer transport study. J. Neurosci. 2:604-622.

Williams, R.W. and K. Herrup. 1988. The control of neuron number. Ann. Rev. Neurosci. 11:423-453.

Williams, R.W., M.J. Bastiani and L.M. Chalupa. 1983. Loss of axons in the cat optic nerve following fetal unilateral enucleation: An electron microscopic analysis. J. Neurosci. 3:133-144.

Williams, R.W., M.J. Bastiani, B. Lia and L.M. Chalupa. 1986. Growth cones, dying axons, and developmental fluctuations in the fiber population of the cat's optic nerve. J. Comp. Neurol. 246:32-69.

Wong, R.O.L. and A. Hughes. 1987a. The morphology, number, and distribution of a large population of confirmed displaced amacrine cells in the adult cat retina. J. Comp. Neurol. 255:159-177.

Wong, R.O.L. and A. Hughes. 1987b. Role of cell death in the topogenesis of neuronal distributions in the developing cat retinal ganglion cell layer. J. Comp. Neurol. 262:496-511.

Activity, Chemoaffinity and Competition: Factors in the Formation of the Retinotectal Map
Ronald L. Meyer

For more than forty years, the retinotectal system of lower vertebrates has been used by developmental neurobiologists as a model to study how growing nerve fibers establish ordered neuronal connections in the central nervous system. For the most part, ordered connections has meant the retinotopic order of the optic projection onto contralateral tectum. Much of the early work has been premised on the assumption that the explanation would come down to a single basic principle, the simpler the hypothesis and the more it could explain, the better.

During the 1970's much of the discussion centered around two extreme hypotheses. One was rigid chemoaffinity in which each neuron was postulated to have a cytochemical identity or marker which rigidly specifies which connections are permissible. At the other extreme was morphogenetics which assumes that neurons have no individual markers and that connectivity is dictated by circumstances external to the cell such as mechanical guidance structures along the pathway and the relative time of origin of afferents and targets.

In spite of a great deal of work and even more discussion neither of these extremes proved adequate as complete explanation. The observations were simply too complicated. Chemoaffinity was confronted by examples of various plasticities in which optic fibers came to innervate foreign territory. Morphogenetics was confronted by ever increasing array of experiments showing the existence of markers. And to make matters worse for both of these, impulse activity has been more recently found to play an important role in the patterning of nerve connections.

The overall view this work leads to is that one simple hypothesis is not likely to explain the data and that one must look to multiple processes. For our lab, the question has now become what are these processes and what does each contribute to the formation of ordered nerve connections. The evidence for the existence of chemoaffinity and activity is strong. However, it is not clear that these are the only factors and evidence that they are not will be presented here. Also our understanding of what chemoaffinity and activity contributes, that is, what aspects of connectivity are regulated by these processes is still rudimentary. Some studies aimed at further clarification of what these two processes contribute will also be described.

CHEMOAFFINITY

Before launching into the actual experiments, it is instructive to review some of earlier data especially as they bear on chemoaffinity and activity. First let us consider the case for chemoaffinity. Perhaps the most fundamental observation was that initially made by Sperry who found that severed optic fibers in lower vertebrates appeared to grow back to their original location in tectum after an eye rotation or other disturbances of the optic pathway (Sperry, 1944; 1945). The observation that regenerating fibers can be forced to follow aberrant paths and yet will reestablish their retinotopic projection has been confirmed with modern methods numerous times and has been extended to initial development. This led Sperry to propose that each retinal ganglion cell and tectal cell possessed position dependent markers and that optic fibers could grow toward their correct target cell using a kind of surface chemotaxis or haptotaxis (Sperry, 1963).

A further step was taken in an experimental design in which part of retina in goldfish was removed and the nerve crushed. Fibers from the remaining retina reinnervated the appropriate part of tectum (Attardi & Sperry, 1963). Comparable findings have been obtained for initial development in chick (Crossland et al, 1974) and hamster (Frost & Schneider, 1979). Similar studies on developing lower vertebrates have been complicated by the capacity of these animals to replace the ablated tissue by regeneration. However, studies in which embryos were examined early in development or those using compound (duplicated) eyes show clear evidence of preferential innervation of tectum (Straznicky et al., 1979; Fawcett & Gaze, 1982). Finally, in goldfish, a piece of tectum can be rotated or translocated to an ectopic tectal position (Yoon, 1973; Gaze & Hope, 1983). When this is done, regenerating optic fibers can track down the appropriate part of tectum to form a rotated or translocated projection.

The seeming fly in the ointment in the case for chemoaffinity is the retinotopic plasticity exhibited by the system. When the posterior half of tectum is removed, optic fibers from the entire retina will squeeze or compress onto the anterior half while preserving retinotopy (Gaze & Sharma, 1970; Yoon, 1976; Meyer, 1977). Or when half of retina is removed, in some species and under certain conditions fibers from the remaining retina will expand across the entire tectum in retinotopic fashion (Schmidt et al., 1978; Meyer, 1987a; Udin & Gaze, 1983). For lower vertebrates where these plasticities are most robustly expressed, these experimental plasticities may be related to the normally occurring shift in the projection that takes place during the extended period of retinal and tectal growth (Gaze et al., 1974; Meyer, 1978b; Cook et al., 1983).

The existence of these retinotopic plasticities should not be taken to mean that the entire notion of chemoaffinity is incorrect. More complex versions of chemoaffinity can and have been proposed to account for many of the observations (Meyer, 1979b; 1982a; Fraser, 1980). Indeed, the existence of markers seems to be the only way to account for the examples of selective innervation and markers need to be invoked to explain the preservation of appropriate polarity exhibited in these plasticities.

ACTIVITY

One finding that is very difficult to explain in terms of chemoaffinity is the eye specific segregation of optic fibers into ocular dominance columns that occurs when two eyes are made to coinnervate a single tectum during development or regeneration in lower vertebrates (Levine & Jacobson, 1975; Constantine-Patton & Law, 1978; Meyer, 1979a). This segregation appears to be actively generated, that is, is not explainable by differences in the initial position or time of ingrowth of fibers. Using the surgical deflection method in goldfish described below, it is possible to make fibers from both left and right eyes regrow into tectum simultaneously and from the same place. Under these conditions, fibers exhibit an initial period of overlap before segregating into columns (Meyer, 1983). During the development of the cat or primate visual cortex, afferents from dorsal lateral geniculate are similarly interdigitated before forming ocular dominance columns (Rakic, 1976; LeVay et al, 1978). A possible way to reconcile this eye by eye segregation with chemoaffinity is to postulate the existence of an eye specific marker such a left-right difference or an idiotypic difference between eyes. This possibility was tested by surgically inducing an embryonic Xenopus eye to internally duplicate in mirror-image fashion so that each half retina projected across the whole tectum. This happens automatically with large ablations of the eye rudiment. By selectively labeling one half of retina, it was shown that fibers from each half of retina segregated into separate ocular dominance columns. Since these fibers originated from single isogenic eyes, a differential marker is quite unlikely (Ide et al., 1983; see also Fawcett & Willshaw, 1982).

In contrast, blocking impulse activity by repeated intraocular injections of tetrodotoxin (TTX) prevents the formation of ocular dominance columns in both lower vertebrates (Meyer, 1982b; Reh & Constantine-Pattton, 1985) and cats (Stryker & Harris, 1986). Of course, ocular dominance columns are not a normal feature of the retinotectal projection and this fact has prompted similar TTX studies during regeneration of one optic nerve in goldfish. Autoradiographic (Meyer, 1983) and electrophysiological (Schmidt & Edwards, 1983) analysis indicated that the projection was not a precisely ordered as normal. It was grossly retinotopic but lacked fine retinotopy.

The now widely offered explanation for these activity sensitive effects is that fibers generate local order on the basis of correlated activity. It is known that neighboring retinal ganglion cells exhibit correlated activity (Arnett, 1978). If fibers followed the simple rule, fibers that fire together terminate together, then retinotopy would be produced. In the case of two eyes coinnervating one tectum, fibers from the same eye would always be more correlated than fibers from the other eye and would aggregate together to generate ocular dominance columns.

MORPHOGENETICS

The evidence that time of fiber ingrowth or mechanical features of the pathway produce order have been equivocal. Much of the evidence is derived from observations on normal animals where detailed analysis of the paths of optic fibers through the nerve and central tracts reveal a high degree of stereotypic organization. However, the organization is complex and may not be due to simple morphogenetics. In fish, for example, there is a reorganization of fibers just central to the optic chiasm which dramatically alters the relative positions of fibers (Scholes, 1979; Fawcett, 1981). This reordering is much easier to explain as marker mediated fiber growth than as passive morphogenetics. Experimental studies are equally problematic. The majority of studies in which the timing and the position of fibers has been perturbed in development or regeneration lead to the conclusion that these don't matter.

A few studies might be argued to show some effect of pathway order on the final projection (Fawcett & Gaze, 1982; Straznicky et al., 1979) but their interpretation is ambiguous. What they really show is that there is some pathway order which is arguably correlated with the distribution of fibers in tectum but there is no evidence that this pathway order is actually causative. The real cause may be chemoaffinity cues in tectum. Furthermore, in many of these studies, the pathway order is appropriate for the retinal origin of the fibers indicating it is generated by chemoaffinity markers within the pathway.

DEFINING CHEMOAFFINITY, ACTIVITY AND MORPHOGENETICS

One of the limitations of previous work is that there is no direct control over fiber path within the target proper. In most experiments the exact path and time of ingrowth of fibers within tectum is unknown. In order to better define the relative contributions of chemoaffinity, activity and morphogenetics, we have devised a method which permits more direct control of the entry point and time of ingrowth of fibers within the tectum itself. This method, referred to as fiber deflection, takes advantage of the orderly path followed by optic fibers within the tectum of normal goldfish and their superficial position in tectum which allows them to be directly visualized in the operating microscope. It is possible to cut away a group of fibers from surrounding tectum, to lift them across the midline and to insert them into the opposite tectum.

This method was used to devise an extreme test of chemoaffinity (Meyer, 1984). Fibers which normally innervate the posterior lateral quadrant of tectum were deflected into the medial anterior quadrant of the opposite "host" tectum and the host eye removed so as to denervate tectum of all other optic fibers (the retinotectal projection is essentially completely crossed). This placed optic fibers as far away from their appropriate part of tectum as possible. It also placed them in the incorrect pathway since optic fibers at that position normally grow posteriorly into medial posterior tectum. Since other optic fibers were absent, deflected fibers could not utilize fiber to fiber interactions to locate their correct part of tectum which has been proposed as an alternative to strict fiber to tectum chemoaffinity.

After several months the deflected fibers were labelled for autoradiography. Deflected fibers were found to have invaded their correct lateral posterior quadrant of tectum where they typically formed their densest projection. However, substantial grains were also found at the insertion site and along the path taken by regenerating fibers into lateral posterior tectum. Electron microscopy of HRP labelled fibers (Hayes & Meyer, 1988) showed these fibers were making substantial numbers of synapses, up to 40% their normal number, near the insertion point in medial anterior tectum. In contrast, when medial posterior fibers were inserted into medial anterior tectum, almost no label could be detected in lateral tectum. Medial fibers instead grew into medial posterior tectum although medial anterior regions were innervated as well .

The finding that fibers grew into their appropriate part of tectum further confirms the existence of a fiber to tectum chemoaffinity. The stability of the putative tectal markers was tested by removing the host eye up to 18 months before deflecting fibers. Selectivity was just as good and, in fact, noticeably better with prior denervation. Impulse activity also made no detectable contribution to the projection as tested by TTX blockade of deflected fibers during regeneration (Meyer, 1987b).

On the other hand, the finding of a substantial medial projection by lateral fibers deflected into medial tectum suggests an effect of the initial starting position. The distribution of label along the apparent growth path (deflected lateral posterior fibers in medial anterior and lateral anterior regions and deflected medial posterior fibers in medial anterior tectum) also indicates an effect of pathway. Both observations can be interpreted to mean that the pathway (morphogenetic factors) can bias the final distribution of fibers.

It might be argued that although the starting position and pathway may have affected the final distribution of fibers, this was only because there were so few fibers. If fiber number were normal, perhaps the formation of an ectopic projection would be prevented. This would mean that pathway factors and differential timing do not matter under normal conditions. There is some reason to believe this. In regeneration following simple nerve crush, some optic fibers do grow into the wrong half of tectum yet do not terminate there. Instead, fibers form a projection that has been often described as being essentially normal (Meyer, 1980; Fujisawa, 1981; Cook et al., 1983).

Since the deflection experiment demonstrated a pathway effect, we wondered if the regenerated projection formed after nerve crush was, in fact, as good as normal. To reexamine the projection more closely, we utilized an anatomical tracing method that was more sensitive to retinotopy than previous methods (Meyer et al, 1985). Small injections of wheat germ agglutinin conjugated to horseradish peroxidase (WGA-HRP) were injected into retina. In normals, this procedure labelled fibers in one small patch of tectum about 200-300 μm wide in tecta which were about 3000 μm wide. The optic nerve was crushed and retina similarly labelled at various times later. At 20 days label was, as expected from previous studies, dispersed over a considerable area of tectum. About a third of tectum contained label. By 40-60 days, label condensed but into 2 or more separate patches. Thus regenerated retinotopy was not only less accurate than normal but qualitatively different.

To confirm this, fine grained electrophysiological mapping was done using an eye-in-water recording apparatus that incorporated a water-filled hemisphere. This permitted 180° of visual field to be mapped with good optics. The projection was not only less retinotopic than normal but different. The multiunit receptive fields which are commonly thought to originate from arbors of optic axons were enlarged from the normal 3° to about 6° as expected from a project with less refined retinotopy. In addition, the positions

of the fields were often retinotopically inappropriate by 20° or
more. In other words, the fields were displaced by a distance that
was substantially larger than the reduced retinotopy indicated by
the enlarged receptive field. This was observed at 6 months or more
after regeneration, generally considered to be quite late in
regeneration (Meyer, unpublished).

There are two salient difference between regeneration and
normal development that might account for the disturbed retinotopy
of the regenerated projection. One is the disordered pathway in
regeneration compared to the much more organized paths of normals.
The other is the virtually simultaneous ingrowth of fibers during
regeneration compared to the sequential ingrowth of fibers during
normal development. It seems unlikely that either pathway order or
timing are directly responsible for producing normal retinotopy in
an instructive sense. Pathway order does not appear to be
sufficiently precise and fibers can, to a considerable extent,
correct pathway errors. Timing also lacks essential information. The
growth of retina is annular which would make for a corresponding
annular ambiguity.

A more plausible interpretation is that the pathway and annular
growth provide permissive conditions for the optimal expression of
the instructive factors, chemoaffinity and activity. In other words,
the arrival of a limited set of fibers in the correct vicinity of
tectum delimits the combinations and permutations that chemoaffinity
and activity need to cope with. In any event, the fact that
retinotopy is not normal following regeneration raises the
possibility of a role, though a limited one, for morphogentic
factors.

Except for this potential contribution by morphogenetics, can
chemoaffinity and activity dependent ordering explain the remaining
data, and, in particular, can they account for retinotopic
plasticity? In principle, they might. Chemoaffinity could specify a
rather low resolution of topography perhaps imparting little more
than an overall appropriate polarity and imposing very weak
constraints on where fibers can terminate in tectum. Activity
dependent ordering could then determine the correct relative
positioning regardless of the absolute position of the fiber in
tectum. Activity could impart the retinotopy exhibited in the
compressed projection of half tecta and in the expanded projection
of half retinas.

To determine if this is the case, activity was eliminated with TTX during the regeneration of fibers onto an anterior half tectum or from a nasal half retina in goldfish (Meyer & Wolcott, 1987). The retinotopy was assayed by electrophysiological mapping after recovery from the last TTX injection at 1-3 months of regeneration. The impulse blockade produced the expected enlargement of multiunit receptive fields to about 20° similar to that previously reported. In spite of this, a retinotopic compression and expansion was clearly evident. The shift in the position of the receptive fields in the part of tectum where plasticity was greatest was as much as $80-90^\circ$, much greater than the enlargement of the receptive fields. Also, plotting the centers of the receptive fields showed the retinotopy to be comparable to half tectum or half retina fish which were not given TTX. This plasticity was confirmed anatomically with autoradiographic mapping methods in which only part of the retina was labelled.

The implication of this last finding is that there is a mechanism for generating retinotopy which is independent of tectal position and does not require activity. What this mechanism is, is presently unknown, but two general possibilities have been pointed out. One has been referred to as "best fit" chemoaffinity. This entails a complex chemoaffinity interaction between fibers and the target in which fibers compete for available tectal sites. As a result of this competition, fibers select their target sites in a context dependent fashion. This context dependent targeting is supposed to produce a match between the competing fibers with the best available target sites such that retinotopy is automatically generated (Meyer, 1982a).

The second possibility is that there is an additional selective interaction, a fiber to fiber interaction such that fibers from the same region of retina congregate together in tectum (Meyer, 1979b; 1982a). This could be mediated by a fiber-fiber contact chemoaffinity. The selective fasciculation of optic fibers from temporal retina with temporal fibers in vitro (Bonhoeffer & Huf, 1985) is suggestive evidence for this and cases of map inversion might be explained in these terms (Meyer, 1979b).

ACTIVITY, ADDITIONAL STUDIES

The preceding studies showing that activity is incapable of explaining retinotopic plasticity and inadequate to generate perfect retinotopy following regeneration raises the issue of what it actually contributes to topography. To reexamine this issue we again mapped the distribution of regenerating optic fibers labelled with small retinal injections of WGA-HRP but this time while blocking with TTX (Olson & Meyer, 1987). As before we found the early projection to be widely dispersed and the later projection to be substantially condensed. This condensation implies that there is retinotopic refinement which does not require activity. Definite proof of this awaits the demonstration that ectopic synapses are made and removed during this condensation.

The condensation under TTX, however, differed from that which occurred with impulse activity. Instead of multiple clumps, there was only one patch of label in tectum and the area encompassed by this patch was several times larger than the total area occupied when there was activity. This implies that activity is responsible for producing the multiple clumps but also indicates that activity operates over a limited range. The distance over which a given fiber can detecte correlated activity is presumably less than the tectal area to which neighboring ganglion cells initially project; otherwise, these fibers should condense into a single spot. What fibers apparently do instead is move toward local hot spots containing fibers with correlated activity which might arise from the differential paths taken by regenerating fibers. The WGA-HRP labelling did, in fact, show multiple concentrations of fibers from one retinal region in early regeneration. Since a similar disorder does not arise in normal development, then activity dependent refinement could lead to neighboring fibers condensing into a single tectal locus.

Since this WGA-HRP retinal injection method was such a sensitive indicator of activity dependent ordering, we used it to address a controversy concerning the existence of a critical period during regeneration. It had been claimed in an electrophysiological study that if activity were eliminated by TTX during the first 35 days of regeneration, map refinement as measured by the enlargement of multiunit receptive fields was permanently prevented (Schmidt & Edwards, 1983). In contrast, the early autoradiographic studies indicated that refinement could be temporarily inhibited with TTX for 40 days but subsequent activity lasting as little as 3 weeks would allow refinement to occur (Meyer, 1983). A similar finding was obtained for ocular dominance columns analyzed anatomically (Meyer, 1982b). A caveat to these early autoradiographic studies was that activity blockade was initiated several weeks into regeneration but we have subsequently confirmed these studies under conditions were TTX blockade was initiated early on and maintained for a longer time.

When activity was blocked for up to 4 months and then activity was allowed to resume for a month or two, spot injections of WGA-HRP into retina produced multiple clumps of label in tectum as had been observed when no TTX was used at all. However, these clumps were typically smaller than those produced with initial activity. This difference suggests a possible explanation for the electrophysiolgical findings. The recording radius of the electrode, the distance over which unit activity can be detected, might be larger than the width of these small clumps. This would mean that both a clump and an adjacent zone innervated by a different region of retina would be simultaneously recorded. As a result, the activity dependent condensation would not be detected electrophysiologically. In this context, it is worth noting that ocular dominance columns in frogs, which are comparable in size to these clumps, went unnoticed for many years because of the inability of the usual electrophysiological mapping methods to resolve them.

The effects of dark rearing may have a similar explanation. Maintaining fish in constant darkness has been reported to inhibit refinement during regeneration when measured electrophysiologically (Schmidt & Eisele, 1985). However, this has not been seen anatomically. Using either autoradiography or the WGA-HRP spot injections, the usual activity dependent refinement was seen (Olson & Meyer, 1987). Ocular dominance columns also formed in darkness. These results are consistent with the cross correlation studies on retina showing neighboring ganglion cells exhibit correlated activity in the dark (Arnett, 1978). This result suggests that the activity pattern responsible for producing these activity dependent effects is "hard wired" into the retinal circuitry and is not easily disturbed with external stimuli. This could explain why strobe illumination also can fail to inhibit refinement and column formation, though some success at inhibiting refinement with strobe has recently been reported (Cook & Rankin, 1986).

These discrepancies between the direct anatomical methods and the electrophysiological recordings points to some of the shortcomings of using electrophysiology to determine afferent distribution. Although it is often argued that the source of the multiunit activity are optic arbors, there is actually no good experimental evidence for this. It is unclear whether the source is presynaptic, postsynaptic or both and how far the electrodes are from the source.

To try to improve our understanding of the functional connectivity formed by optic fibers during regeneration, we decided to record from single postsynaptic tectal cells during regeneration (Brink and Meyer, 1988). We found that we could reliably record a population of visually driven intrinsic tectal neurons located in the lower half of the main optic innervation layer (SFGS). These were determined to be postsynaptic based on their response to optic nerve shock and had small receptive fields that would make them ideal for monitoring during regeneration.

We were diverted from this aim, however, by some particularly interesting features of spontaneous tectal activity. In the absence of visual stimulation these cells exhibited relatively little spontaneous activity. When the nerve was crushed, tectal neurons showed considerable spontaneous activity that was notable for its phasic nature. This "bursting" was particularly evident in multiunit recordings, that is, the bursting was comprised of different units. Accompanying these bursts were periodic episodes of DC potentials which were several millivolts negative and several tens of milliseconds long. This pattern of activity was detectable beginning at a few days after nerve crush until fibers began to reinnervate tectum at around 3 weeks. If the eye was removed, the activity persisted for at least 3 months.

To distinguish whether this periodic activity represented phasic activity of the entire tectum or locally correlated activity, simultaneous recordings from denervated tecta were done with two electrodes and the activity analyzed with cross-correlation methods. When the two electrodes were 100-200 μm apart, the two recorded trains showed a strong positive correlation having a peak with a base about 50-75 μm wide. At 300 μm, the correlation was substantially less and at about 400 μm little or no correlation was detectable. Thus, the spontaneous activity was locally but not globally correlated.

The existence of local coupling between tectal cells suggests an answer to an apparent quandary over how activity dependent refinement and ocular dominance columns are supposed to form. It is widely presumed that the underlying mechanism is a Hebbian synapse. According to Stent's (1973) modern formulation of this mechanism, those synapses which are active when the postsynaptic cell fires are favored (stabilized) while those synapses which were silent when the postsynaptic cell fires are disfavored (destabilized). This simple Hebbian synapse can readily explain how two coactive fibers can come to selectively innervate the same target cell. Unfortunately, it also predicts a cell by cell segregation of afferents and this is not what is observed. Fibers from one region of retina or, in the case of ocular dominance columns, fibers from one eye preferentially innervate a local region of tectum which comprises many target cells.

A simple extension of the Hebbian synapse can accommodate the results. If target cells are locally coupled and the Hebbian synapse can operate over tens of milliseconds, then the fibers might treat all the coactive tectal cells as if they were one cell. The target for activity dependent sorting would thus be the group of coactive target cells not the individual cells. The extent and the pattern of coupling would presumably determine important features of the projection.

QUANTITATIVE AND QUALITATIVE SYNAPTIC SPECIFICITY

The preceding studies related to the effects of activity point to the synapse as the crucial nexus for mediating the formation of activity dependent order. This prompted a series of ultrastructural studies into the normal pattern of synapses formed by optic fibers in tectum and into their synaptogenesis during regeneration. One of our aims was to sort out which ultrastructural features may be regulated by activity from those that are determined by the intrinsic properties of the cells.

A first step was to simply quantitate the number of synapses that optic fibers formed in tectum during regeneration with or without impulse blockade. There were a number of issues that might be resolved by a definitive study of this kind. One was whether synaptic number is strongly regulated during synaptogenesis. Another was whether activity is responsible for regulating synapse number.

Such a study might also distinguish between highly selective synapse formation versus random synapse formation followed by synapse elimination or rearrangement.

We were not the first to ask some of these questions but previous efforts had produced conflicting results (Murray & Edwards, 1982; Murray et al., 1982; Marotte, 1983). There were two technical difficulties with previous studies that cast doubts about these previous studies. One was the problem of identifying optic synapses in a manner suitable for quantitation. Although it was possible to label normal optic fibers with HRP, it had proven very difficult to label regenerating fibers possibly because of their small diameter. As as a result, previous studies were forced to count all synapses and infer the proportion that were optic or used degeneration to identify optic synapses. The other technical problem was the sampling method. The main optic innervation layer of tectum exhibits pronounced changes in thickness upon denervation and subsequent reinnervation. The usual random sampling procedures require ad hoc assumptions about the depth distribution of optic synapses.

To overcome these limitations, William Hayes and I (Hayes & Meyer, 1985; 1986; 1988) developed a method for labelling regenerating optic fibers for quantitative ultrastructure and a sampling procedure that did not require arbitrary assumptions about distribution of optic synapses. To label fibers, the eye was removed and the cut end of the optic nerve sucked up into a small tube filled with concentrated HRP. The fish was then placed in the refrigerator which appeared to both retard degeneration and aid filling. The tectum was then prepared using standard EM histochemical methods. To count synapses, a single thin section was taken through the entire tectum which is only a few hundred micrometer thick. From this section, a series of overlapping electron micrographs were taken and assembled into a montage that spanned the entire SFGS, the main optic innervation lamina, and the adjacent non-optic layers. All optic synapses and fibers within this column or "core sample" were counted and their depth distribution plotted. Since this tangential dimension (width and length) of tectum is known to be constant (Murray & Edwards, 1982), this sampling procedure would give an accurate comparison of the relative number of optic synapses.

Using this column sampling procedure, it was found that in normals about 43% of the total synapses in the SFGS were optic but that their density varied as a function of depth exhibiting a pronounced low in a band just above the bottom of the SFGS. The counts were repeated in fish with regenerating optic fibers at 30, 60 and 240 days. Thirty days corresponds to the time when fibers have reinvaded the entire tectum but before the major period of activity dependent refinement which is largely complete by sixty days. The 240 day time point was chosen for comparison as late regeneration.

Contrary to the previous suggestion that synapses would slowly increase in number (Murray and Edwards, 1982), we found that the number of optic synapses were essentially normal by 30 days and remained constant at 60 and 240 days. This result implies that activity dependent ordering is accomplished by synaptic rearrangement of a fixed number of synapses rather than by synapse elimination or highly selective synaptogenesis. Additional evidence for rearrangement came from counts of optic fibers. The number of optic fibers in the sample column at 30 days was about ten times higher than in normals indicating extensive sprouting of the optic axons. By 60 days this number was reduced by half indicating a dramatic restructuring of the fibers.

This same analysis was again repeated for 30 and 60 days regeneration but this time under continuous impulse blockade by repeated intraocular injections of TTX (Hayes & Meyer, 1986). One might expect that fewer synapses might be formed since in cat and primate visual cortex unilateral eye occlusion leads to a reduction in the space occupied by the "deprived" lateral geniculate fibers (Hubel et al., 1977; Shatz & Stryker, 1978). One the other hand one might expect more synapses since activity is thought to be responsible for synapse elimination following the formation of exuberant synapses in other systems. Instead, we found the synaptic counts were virtually identical to those obtained in fish with impulse activity. This result indicates that activity played no role in regulating the number of synapses. We also found, to our surprise, that the numbers of sampled optic fibers were also unaffected, that is, the usual increase in sprouts at 30 days was followed by their usual reduction by 60 days. This implies that the restructuring of optic fibers during regeneration is not produced by activity. However, it also suggests why the effects of activity seem to be so pronounced between 30 and 60 days. This corresponds to a period of automatic (not activity dependent) restructuring of fibers so that any process such as activity which can modulate the distribution of synapses during such restructuring will have its greatest impact during this period. To use an analogy, the faster the car is moving, the greater the effect of a change in steering.

Given that the number of optic synapses is tightly regulated by the system, where is the regulation? Is it that each optic fiber can only make a fixed number of synapses or is that each tectal target neuron only permits a certain number of synaptic sites? A way to decide this issue is offered by the capacity of optic fibers to form a compressed projection onto a surgically formed anterior half tectum. In this case, the normal number of fibers project onto half the normal amount of target. If the total number of optic synapses in the half tectum is normal then regulation lies in the fibers. If the number is half normal, then it lies with the tectal cell.

Since previous attempts at counting the half tectum came to exactly opposite conclusions (Murray et al., 1982; Marotte, 1983), we reexamined the issue using direct labelling of optic fibers and our column sampling method (Hayes & Meyer, 1987). We found that there was the normal number of optic synapses in each sample column. Since these sample columns were of fixed width, this meant that, on average, each tectal target neuron was receiving its normal complement of optic synapses while each optic fiber was forming half as its normal number of synapses. We also found that the depth of the sample column over which optic fibers were distributed was increased by about 50%, that is, the innervation layer was greatly hypertrophied. This may have been due to a large increase in the number of optic fibers per unit area of tectum. The consequence of this hypertrophy was that the density of optic synapses was actually decreased. Thus, it is absolute number of synapses per target that is regulated, not density.

What would happen under the opposite conditions in which a reduced number of fibers are made to innervate an intact tectum? This condition was accomplished using the deflection method by redirecting a select fraction of fibers constituting about 20% of the normal number into the opposite tectum which was denervated of other optic fibers by removing its corresponding eye (Hayes & Meyer, 1988). The result of the counts was that the sample columns all contained 40% or less of the normal number of optic fibers. This meant that each innervated tectal cell, on average, received fewer synapses. Thus, there are limits as to how many synapses a fiber can support. This result is hardly surprising since there are undoubtedly limits on the biosynthetic capacity of the ganglion cells. What was not predictable was that subnormal numbers of synapses were found in all sample columns even though the columns were taken from parts of tectum that were among the most heavily labelled. This suggests that fibers skip potential synaptic sites when their density is low. Another conclusion from this study is that fibers can readily form substantial numbers of synapses at retinotopically incorrect positions. When lateral posterior fibers were inserted into medial anterior tectum, up to 40% the normal number of synapses were found near the insertion site, that is, about as far away from the appropriate tectal quadrant as it was possible to be.

The preceding studies address the issue of quantitative specificity, numbers of synapses, but, for the most part, they leave open the question of qualitative specificity. Are optic fibers synapsing on the correct type of cell at all times during regeneration? What factors regulate their choice of target? To try to address these issues, we have turned to histochemistry and immunohistochemistry to look for markers that would be useful for further distinguishing target cells and different classes of optic fibers at both the light and electron microscope level.

One marker that has proven useful for distinguishing different parts of the CNS of mammals is cytochrome oxidase (CO) which produces a stain that can be considered to be an index of the level of oxidative metabolism. An added bonus of CO is that different levels of CO reactivity have been correlated with differences in impulse activity and so provides an index of differential activity (Wong-Riley & Riley, 1983). Using CO histochemistry to stain tectum (Kageyama & Meyer, 1985; 1988), we found that at the light level the normal tectum is differentiated into a series of lamina of quite different reactivity. The most intense reactivity was localized to the optic laminae, especially the SFGS. This reactivity was lost within 2-3 days after optic denervation indicating that the reactivity was directly associated with the optic input.

There was one notable non-optic but moderately reactive band that was unaffected by optic denervation. It was located immediately subadjacent to the SFGS at the top of the Stratum Grisem Centrale (SGC) which we have designated as SGCa. This band provides, for the first time, a clear marker for the bottom the SFGS and allowed us to address an issue related to the laminar plasticity mentioned above. The thickness of the apparent main optic innervation layer changed dramatically during regeneration, shrinking upon denervation and greatly expanding upon reinnervation, particularly in the case of compression onto a half tectum. Although it was reasonably clear that the upper boundary of SFGS was not changing, the absence of markers for the lower limit of the SFGS made it impossible to tell whether the increased thickness was due to a real expansion of the SFGS or an invasion of optic fibers into the subadjacent non-optic layers. Using the CO staining of SGCa in combination with simultaneous labelling of optic fibers with HRP or cobaltous-lysine, Kageyama and I were able to show that optic fibers remained confined to the SFGS during regeneration onto a half tectum. Essentially all of the thickness changes of tectum following denervation and reinnervation were confined to the optic lamina. Since most tectal cells that have dendrites in the SFGS also send dendrites to other laminae (Meek, 1983), these dimensional changes imply regional dendritic remodelling during that denervation and reinnervation.

Some evidence for cellular specificity was also obtained using CO histochemistry by closely examining the SFGS at the light and EM levels. A number of CO features were associated with specific lamina. One the most salient was a class of very large and highly reactive optic terminals at the very bottom of the SFGS. Using a recently developed pre-fixation HRP method in which the brain was first fixed with EM fixative and then HRP applied to the cut nerve, we were able to preferentially fill large optic fibers with short fill times. With anterograde filling the large terminals in deep SFGS filled and with retrograde filling the large ganglion cells in retina filled suggesting that these terminals derived from large ganglion cells. When the optic nerve was crushed, these large terminals reappeared in this same layer.

A quite unexpected finding in the CO study was the slow recovery of CO reactivity in the optic lamina during regeneration (Kageyama & Meyer, 1985). At thirty days when the normal numbers of synapses were reformed, reactivity was only moderately elevated above the low level associated with complete optic denervation. Several months were required for recovery of normal reactivity and this was associated with the reformation of sublaminar CO banding. This slow recovery was not due to an absence of mitochondria in the optic fibers. Mitochondria were present but they were much less reactive than seen in normal optic fibers. The retinal ganglion cells themselves exhibited elevated reactivity so this did not represent an inability of ganglion to produce reactive mitochondria.

The only good correlation that we could find is with the slow refinement of retinotopy which follows a similar time course. This refinement can be expected to lead to the highest convergence of coactive optic fibers and hence the greatest ionic activity. Ion pumping has been proposed as the best overall explanation for elevated CO reactivity in a variety of other systems. If true here, it suggests that the most refined retinotopic order corresponds to the most elevated ionic pumping and the most intense oxidative metabolism in terminals and dendrites.

Immunocytochemistry offers the potential for identifying specific types of cells so we have turned to immunological methods with the hope of defining connectivity at the cellular level. Some of our currents efforts have utilized monoclonal methodology to generate cell specific markers. A number of lamina specific and tectal cell specific antibodies have been isolated, but these have not yet been characterized and will not be described here. Most of our current effort is focused on transmitter specific antibodies for several reasons. The antigens are known. Transmitters are likely to be pivotal in mediating, at some level or another, the activity dependent ordering. Some of the immunogens are glutaraldehyde linked and consequently survive glutaraldehyde fixation thereby permitting excellent ultrastructure. Two of these glutaraldehyde tolerant anti-transmitters will be described, glutamate and GABA.

Polyclonal antibodies directed against glutaraldehyde conjugated glutamate or aspartate were generated and affinity purified by Bob Wenthold at NIH. Dot blot analysis and preabsorbtion controls on tissue sections indicate the antibodies to be highly selective for glutamate and aspartate. When this anti-glutamate was applied to tectal sections (Kageyama & Meyer, 1987), both cell body and fiber staining was observed throughout most tectal lamina, but the fiber staining was particularly notable in the main optic innervation layer, SFGS. Similar fiber staining was observed in the optic nerve. In both cases, the reactive fibers were dispersed as discreet bundles that were clearly separated from each other suggesting that not all optic fibers were staining. When the eye was removed, this fiber staining in the nerve and SFGS disappeared within a few days. At the EM level, immunoreactivity was localized to synaptic vesicles of terminals in the SFGS, some of which appeared to be optic based on mitochondrial morphology. Cross sections through retina revealed the expected localization in rods and bipolar cells and, in addition, showed strong staining in some retinal ganglion cells. In tangential sections through the ganglion cell layer, it was evident that it was mainly the large ganglion cells which comprise less than 10% of retina that were heavily stained. Anti-aspartate showed a similar but not quite identical staining pattern in retina and tectum. Thus, some but perhaps not all retinal ganglion cells are glutaminergic.

Anti-GABA reactivity was dispersed throughout the tectum (Kageyama & Meyer, 1987) but was clearly distinguishable from anti-glutamate. Only cells were stained and these were smaller and in some cases differently positioned than those showing anti-glutamate reactivity. Some of the reactive cells were in the SFGS. At the EM level, reactivity was again localized to synaptic vesicles. Some of these vesicle were in presynaptic dendrites which were observed making synaptic contacts onto other unreactive dendrites. These dendrites containing reactive vesicles were also postsynaptic to fibers presumed to be optic on the basis of mitochondrial criteria. From previous studies using HRP labelling of optic fibers, it is known that optic fibers do synapse upon vesicle containing profiles which in turn form a serial or triadic synapses upon other dendrites. Thus one of the targets of optic fibers appears to be inhibitory cells which form synaptic patterns similar to those reported for mammalian lateral geniculate and colliculus. How these synaptic patterns change upon denervation and reinnervation is currently under study.

PHARMACOLOGIC CHARACTERIZATION

Another approach to the connectivity question which complements the transmitter immunohistochemistry is pharmacologic analysis. As a step in this direction, we have been analyzing optic transmission using an in vitro system (Van Deusen & Meyer, in preparation) similar to that used to study brain slices (Teyler et al., 1981). Like brain slices, tectum is placed in a recording chamber through which oxygenated Ringer's solution is circulated but with a significant difference: the tectum is not sliced. Tectum can be easily removed from the rest of the brain as a single structure due mainly to the large third ventricle which underlies the tectum. This leaves the intrinsic tectal connections intact, thereby reducing slicing artifact. In addition, the entire optic tract and optic nerve can be removed along with the tectum, making it possible to stimulate optic fibers with high selectivity and high current.

Although single units can be recorded in this preparation, we have to date mainly used the field potential evoked by optic nerve shock since this provides a much larger sample of optic fiber transmission. We initially examined the effects of nicotinic antagonists because of previous in vivo studies claiming that optic transmission in the frog and goldfish was nicotinic cholinergic (Freeman et al., 1980; Schmidt & Freeman, 1980). Using the in vitro method where the concentration of drug is under more direct control, we found no inhibitory effect of curare on the field potential contraindicating nicotinic cholinergic transmission. Instead curare enhanced a late component of the field potential indicating that the antagonist was effective but at some downstream synapses.

In contrast, some glutamate antagonists had a pronounced inhibitory effect on the entire field potential. Kynurenic acid, a broad spectrum antagonist effective at NMDA, quisqualate and kainate receptor subtypes, produced about a 50% decrement of the potential. The specific NMDA antagonist, APV, showed little or no effect. This suggests that optic fibers utilize glutamate at either quisqualate or kainate receptors in tectum. However, not all optic fibers may be glutamanergic. Even at very high concentrations of kynurenic acid (1mM) a substantial field potential was observed even though we could show that only optic fibers were being stimulated. This partial blockade was apparently not due to limited penetration by the drug since another projection to tectum, the torus longitudinalis, could be completely blocked at this concentration of kynurenate in spite of the fact these synapses were located further from the bath than the optic synapses. This partial blockade fits with the anti-glutamate immunohistochemistry in indicating that only some optic fibers are glutaminergic.

CONCLUSION

Although we still cannot claim to completely understand how optic fibers form a retinotopic projection even in broad general terms, we can finger several distinct processes and suggest what they might be contributing to retinotectal order. Some kind of chemoaffinity is clearly important for global organization, overall polarity and gross retinotopy. As Sperry envisioned, this is mediated by position dependent markers in both retina and tectum and a differential response of fibers from different retinal regions to different tectal positions. However, the nature of this response is more complex than a simple homing response of individual fibers toward their "correct" tectal position. The response may instead be vectorial (Meyer, 1984), that is, local cues may impart direction and speed. Fibers can terminate anywhere but their probability of doing so is inversely proportional to their rate of translocation which is highest the farther fibers are from their appropriate position. The response of fibers to tectal cues may also be dependent on the presence of other optic fibers. These interfiber interactions may be responsible for the intermediate range order expressed experimentally in compression and expansion and normally during shifting connections that accompany growth. There is evidence for two kinds of interfiber interactions, fasciculation and competition for synaptic sites. Which, if either, of these is responsible for retinotopic plasticity is unknown. Activity is important for short range order but this order may turn out to be more complicated than refined retinotopy. The potential involvement of linked groups of tectal cells opens the possibility that activity may be responsible for the assembly of specific kinds of circuits.

This perspective, that multiple processes contribute distinct features to the organization of connections, is critically important for the future search for the biochemical events that underly the formation of ordered nerve connections. An effective strategy will probably require a focused attack on each process coupled with a clear understanding of what that process contributes.

REFERENCES

Arnett, D.W. 1978. Statistical dependence between neighboring retinal ganglion cells in goldfish. Exp. Brain Res. 32:49-53.

Attardi, D.G., and R.W. Sperry. 1963. Preferential selection of central pathways by regenerating optic fibers. Exp. Neurol. 7:46-64.

Bonhoeffer, F., and J. Huf. 1985. Position-dependent properties of retinal axons and their growth cones. Nature 315:409-410.

Constantine-Paton, M., and M.I. Law. 1978. Eye-specific termination bands in tecta of three-eyed frogs. Science 202:639-641.

Cook, J.E., A.J. Pilgrim, and T.J. Horder. 1983. Consequences of misrouting goldfish optic axons. Exp. Neurol. 79:830-844.

Cook, J.E., and E.C. Rankin. 1986. Impaired refinement of the regenerated retinotectal projection of the goldfish in stroboscopic light: a quantitative WGA-HRP study. Exp. Brain Res. 63:421-430.

Cook, J.E., E.C. Rankin, and H.P. Stevens. 1983. A pattern of optic axons in the normal goldfish tectum consistent with the caudal migration of optic terminals during development. Exp. Brain Res. 52:147-151.

Crossland, W.J., W.M. Cownan, L.A. Rogers, and J.P. Kelly. 1974. The specification of the retino-tectal projection in the chick. J. Comp. Neurol. 155:127-164.

Fawcett, J.W. 1981. How axons grow down the Xenopus optic nerve. J. Embryol. Exp. Morphol. 65:219-233.

Fawcett, J.W., and R.M. Gaze. 1982. The retinotectal fibre pathways from normal and compound eyes in Xenopus. J. Embryol. Exp. Morph. 72:19-37.

Fraser, S.E. 1980. A differential adhesion approach to the patterning of nerve connections. Dev. Biol. 79:453-464.

Freeman, J.A., J.T. Schmidt, and R.E. Oswald. 1980. Effect of a-bungarotoxin on retinotectal synaptic transmission in the goldfish and the toad. Neurosci. 5:929-942.

Frost, D.O., and G.E. Schneider. 1979. Plasticity of retinofugal projections after partial lesions of the retina in newborn syrian hamster. J. Comp. Neurol. 185:517-568.

Fujisawa, H.. 1981. Retinotopic analysis of fiber pathways in the regenerating retinotectal system of the adult newt Cynops Pyrrhogaster. Brain Res. 206:27-37.

Gaze, R.M., and R.A. Hope. 1983. The visuotectal projection following translocation of grafts within an optic tectum in the goldfish. J. Physiol. 344: 257-275.

Gaze, R.M., M.J. Keating, and S.H. Chung. 1974. The evolution of the retinotectal map during development in Xenopus. Proc. R. Soc. Lond. B. 185:301-330.

Gaze, R.M., and S.C. Sharma. 1970. Axial differences in the reinnervation of the goldfish optic tectumby regeneration optic nerve fibers. Exp. Brain Res. 10: 171-181.

Hayes, W.P., and R.L. Meyer. 1985. Clustered synapse formation by early regenerating retinotectal fibers precedes the sublaminar redeployment of retinal connections. Soc. Neurosci. 11:977.

Hayes, W.P., and R.L. Meyer. 1986. Retinotectal synapse numbers are regulated by an activity- independent and target-dependent mechanism in goldfish. Soc. Neurosci. 12:436.

Hayes, W.P., and R.L. Meyer. 1988. Retinotopically inappropriate synapses of subnormal density formed by surgically misdirected optic fibers in goldfish tectum. Dev. Brain Res. 38:304-312.

Horder, T.J., and K.A.C. Martin. 1978. Morphogenetics as an alternative to chemospecificity in the formation of nerve connections. In A.S.G. Curtis (ed): Cell-cell recognition. Cambridge: Cambridge Univ. Press, pp. 275-358. .pa

Hubel, D.H., T.N. Wiesel, and S. LeVay. 1977. Plasticity of ocular dominance columns in monkey striate cortex. Phil. Trans. Roy. Soc. London B.278:377-409.

Ide, C.F., S.E. Fraser, and R.L. Meyer. 1983. Eye dominance columns from an isogenic double-nasal frog eye. Science 221:293-295.

Kageyama, G.H., and R.L. Meyer. 1985. Histochemical localization of cytochrome oxidase in the normal and denervated goldfish optic tectum: a combined golgi-cytochrome oxidase study. Soc. Neurosci. 11:236.

Kageyama, G.H., and R.L. Meyer. 1987. Immunohistochemical localization of gaba, choline acetyltransferase, glutamate and aspartate in the visual systems of goldfish and mice. Soc. Neurosci. 13:860.

Kageyama, G.H., and R.L. Meyer. 1988. Histochemical localization of cytochrome oxidase in the normal retina and optic tectum of goldfish: A combined C.O. - HRP study. J. Comp. Neurol. :.

LeVay, S., M.P. Stryker, and C.J. Shatz. 1978. Ocular dominance columns and their development in layer IV of the cat's visual cortex. J. Comp. Neurol. 179:223-244.

Levine, R.L., and M. Jacobson. 1975. Discontinuous mapping of retina onto tectum innervated by both eyes. Brain Res. 98:172-176.

Marotte, L.R.. 1983. Increase in synaptic sites in goldfish tectum after partial tectal ablation. Neurosci. Lett. 36:261.

Meek, J.. 1983. Functional anatomy of the tectum mesencephali of the goldfish: An explorative analysis of the functional implication of the laminor structure organization of the tectum. Brain Res. Rev. 6:247-297.

Meyer. 1983. Tetrodotoxin inhibits the formation of refined retinotopography in goldfish. Brain Res. 6:293-298.

Meyer, R.L.. 1977. Eye-in-water electrophysiological mapping of goldfish with and without tectal lesions. Exp. Neurol. 56:23-41.

Meyer, R.L.. 1978a. Deflection of selected optic fibers into a denervated tectum in goldfish. Brain Res. 155:213-227.

Meyer, R.L.. 1978b. Evidence from thymidine labelling for continuing growth of retina and tectum in juvenile goldfish. Exp. Neurol. 59:99-111.

Meyer, R.L.. 1979a. "Extra" optic fibers exclude normal fibers from tectal regions in goldfish. J. Comp. Neurol. 183:883-902.

Meyer, R.L.. 1979b. Retinotectal profjection in goldfish to an inappropriate region with a reversal in polarity. Science 205:819-821.

Meyer, R.L.. 1980. Mapping the normal and regenerating retinotectal projection of goldfish with autoradiographic methods. J. Compl Neurol. 189:273-289.

Meyer, R.L.. 1981. "Ocular dominance" columns in goldfish, ontogeny and effect of visual environment. Soc. Neurosci. 7:405.

Meyer, R.L.. 1982a. Ordering of retinotectal connections: a multivariate operational analysis. Curr. Top. Dev. Biol. 17:101-145.

Meyer, R.L.. 1982b. Tetrodotoxin blocks the formation of ocular dominance columns in goldfish. Science 218:589-591.

Meyer, R.L.. 1983. The growth and formation of ocular dominance columns by deflected optic fibers in goldfish. Dev. Brain Res. 6:279-291.

Meyer, R.L.. 1984. Target selection by surgically misdirected optic fibers in the tectum of goldfish. J. Neurosci. 4:234-250.

Meyer, R.L.. 1987a. Tests for relabelling the goldfish tectum by optic fibers. Dev. Brain Res. 31: 312-318.

Meyer, R.L.. 1987b. Intratectal targeting by optic fibers in goldfish under impulse blockade. Dev. Brain Res. 37:115-124.

Meyer, R.L. and D.L. Brink. 1988. Locally correlated activity in the goldfish tectum in the absence of optic innervation. Dev. Brain Res. in press.

Meyer, R.L., K. Sakurai, and E. Schauwecker. 1985. Topography of regenerating optic fibers in goldfish traced with local wheat germ injections into retina: Evidence for discontinuous microtopography in the retinotectal projection. J. Comp. Neurol. 239:27-43.

Meyer, R.L., and L.L. Wolcott. 1987. Compression and expansion without impulse activity in the retinotectal projection of goldfish. J. Neurobiol. 18:549-567.

Murray, M., and M.A. Edwards. 1982. A quantitative study of the reinnervation of the goldfish optic tectum following optic nerve crush. J. Comp. Neurol. 209: 363-373.

Murray, M., S. Sharma, and M.A. Edwards. 1982. Target regulation of synaptic number in the compressed retinotectalprojection of goldfish. J. Comp. Neurol. 209:374-385.

Olson, M.D., and R.L. Meyer. 1987. Refinement of the goldfish retinotectal projection in the absence of activity and in the dark. Soc. Neurosci. 13:1418.

Rakic, P.. 1977. Prenatal development of the visual system in rhesus monkey. Phil. Trans. R. Soc. Lond. B. 278:245-260.

Reh, T., and M. Constantine-Paton. 1985. Eye-specific segregation requires neural activity in three-eyed Rana pipiens. J. Neurosci. 5:1132-1143.

Schmidt, J.T., C.M. Cicerone, and S.S. Easter. 1978. Expansion of the half retinal projection to the tectum in goldfish:an electrophysiological and anatomical study. J. Comp. Neurol. 177:257-278.

Schmidt, J.T., and D.L. Edwards. 1983. Activity sharpens the map during the regeneration of the retinotectal projection in goldfish. Brain Res. 269:29-39.

Schmidt, J.T., and L.E. Eisele. 1985. Stroboscopic illumination and dark-rearing block the sharpening of the regenerated retinotectal map in goldfish. Neuroscience 14:535-546.

Schmidt, J.T., and J.A. Freeman. 1980. Electrophysiological evidence that retinotectal synaptic transmission in the goldfish is nicotinic cholinergic. Brain Res. 187:129-142.

Scholes, J.H.. 1979. Nerve fiber topography in the retinal projection to the tectum. Nature 278:620-624.

Shatz, C.J., and M.P. Stryker. 1978. Ocular dominance in layer IV of the cat's visual cortex and the effects of monocular deprivation. J. Physiol. 281:267-283.

Sperry, R.W.. 1944. Optic nerve regeneration with return of vision in anurans. J. Neurophysiol. 7:57-69.

Sperry, R.W.. 1945. Restoration of vision after crossing of optic nerves and after contralateral transposition of the eye. J. Neurophysiol. 8:15-28.

Sperry, R.W.. 1963. Chemoaffinity in the orderly growth of nerve fiber patterns and connections. Proc. Nat. Acad. Sci. U.S.A. 50:703-710.

Stent, G.S.. 1973. A physiological mechanism for Hebb's postulate of learning. Proc. Nat. Acad. Sci. U.S.A. 70:997-1001.

Straznicky, C., R.M. Gaze, and T.J. Horder. 1979. Selection of appropriate medial branch of the optic tract by fibres of ventral retinal origin during development and in regeneration: An autoradiographic study in Xenopus. J. Embryol. Exp. Morph. 50:253-267.

Stryker, M.P., and W.A. Harris. 1986. Binocular impulse blockade prevents the formation of ocular dominance columns in cat visual cortex. J. Neurosci. 6:2117-2133.

Teyler, T.J., D. Lewis, and V.E. Shashoua. 1981. Neurophysiological and biochemical properties of the goldfish optictectum maintained in vitro. Brain Res. Bull. 7:45-56.

Udin, S.B., and R.M. Gaze. 1983. Expansion and retinotopic order in the goldfish retinotectal map after large retinal lesions. Exp. Brain Res. 50:347-352.

Wong-Riley, M., and D.A. Riley. 1983. The effect of impulse blockage on cytochrome oxidase activity in the cat visual system. Brain Res. 261:185-193.

Yoon, M.. 1973. Retention of the original topographic polarity by the 1800 rotated tectal reimplant in young goldfish. J. Physiol. 233:575-588.

Yoon, M.. 1976. Progress of topographic regulation of the visual projection in the halved optic tectum of adult goldfish. J. Physiol. 257:621-643.

Neuronal Surface Receptors in Axon Fasciculation and Regeneration

Vance Lemmon, Kathryn Farr, and Carl Lagenaur

During the development and regeneration of the nervous system, cell adhesion molecules have been implicated in the guidance of migrating neurons and the routing of growing axons. Recent advances have been made in understanding the molecular basis for neuron-neuron and neuron-glial adhesion. These findings have raised new questions concerning the respective contributions to axon outgrowth that are made by the different identified cell adhesion molecules. We have been studying how two cell adhesion molecules, the 8D9 antigen (which is similar or identical to the glycoproteins described by other investigators as NILE, L1, and Ng-CAM) and N-CAM function in neuron-neuron and neuron-glial interactions that result in neurite extension. We have evidence that indicates that the 8D9 antigen plays a vital role in neurite outgrowth and could play an essential part in promoting both peripheral and central nervous system regeneration. N-CAM appears to be a less effective substrate but the function of N-CAM in this process may be heavily dependant on the precise form of the N-CAM.

At the present time, several different molecules have been implicated in adhesion of neurons to other cells and extracellular substrates. These include N-CAM/BSP-2/D2 (Edelman, 1985; Keilhauer et al., 1985; Nobel et al., 1985), NILE/L1/Ng-CAM/8D9 (Stallcup and Beasley, 1985; Rathjen and Schachner, 1984; Grumet et al., 1983; Lemmon and McLoon, 1986), J1 (Kruse et al., 1985), N-cadherin (Hatta and Takeichi, 1986) and myelin-associated glycoprotein (MAG) (Kruse et al., 1985). Laminin (Smalheiser et al., 1984; Bozyczko & Horwitz, 1986) and heparan sulfate (Cole et al., 1985), have also been found to be functional adhesive substrates for neurons. A few other molecules that may participate in nerve-substrate adhesion, such as polyornithine-binding neurite-promoting factor (PNPF) (Davis et al., 1985), seem to represent a class of laminin-like molecules (Lander et al., 1985).

Evidence from a variety of laboratories suggest that molecules related to the 8D9 antigen are associated with neurite outgrowth and cell adhesion. We will present evidence below that the purified 8D9 antigen functions as a substrate for the outgrowth of axons. The identification of this activity in association with the isolated 8D9 antigen suggests that its role in development and regeneration of the nervous system is to provide a pathway for rapid axon outgrowth along already existing axons or glia (Schwann cells). The molecules discussed below, NILE, L1, Ng-CAM, and 8D9 are all of similar molecular weight and are all immunologically cross reactive; their anatomical distribution and developmental expression in the nervous system is similar in the respective species that have been studied (Friedlander et al., 1986; Bock et al., 1985; Lemmon & McLoon, 1986).

McGuire and co-workers first described the NILE (NGF inducible large external) glycoprotein on rat PC12 pheochromocytoma cells (McGuire et al., 1978). Studies with neural cells in vitro indicated that NILE was expressed by all types of neurons as well as Schwann cells but was not found on chromaffin cells or cerebellar glia (Stallcup et al., 1983). These workers suggested that NILE might be involved in neurite extension, because NILE was induced when cells were not grown attached to collagen (and hence unable to extend neurites). NILE was subsequently implicated in neuron-neuron adhesion (neurite fasciculation) (Stallcup & Beasly, 1985).

Schachner and co-workers described a molecule from mouse brain, L1 (Lindner et al., 1983; Rathjen & Schachner, 1984), which they believed was important in neuron-neuron adhesion and neuron migration, but not neuron-glial adhesion (Keilhauer et al., 1985). L1 and NILE have been demonstrated to be immunologically related molecules (Bock et al., 1985). Like NILE, L1 was detected on neurons but not on astrocytes, oligodendrocytes or fibroblasts (Rathjen and Schachner, 1984).

Ng-CAM, was first identified by Edleman and co-workers (Grumet et al., 1983) in chick brain. They found this molecule using an adhesion assay that used brain derived flat cells as a substrate and brain cells as probe cells. Since their flat cells were derived predominantly from glia and their probe cells were predominantly neurons, they proposed that Ng-CAM mediated neuron-glial adhesion. Subsequently, this group has demonstrated that Ng-CAM is also involved in neuron-neuron adhesion and is immunologically related to NILE (Friedlander et al., 1986).

We have identified a molecule in the chick retina, the 8D9 antigen, which we have demonstrated to be immunologically related to L1 (Lemmon & McLoon, 1986; see below). Our studies of the developmental expression of the 8D9 antigen in the chick primary visual pathway have suggested a role in fasciculated axon outgrowth.

Based on the above work, it seems clear that the 8D9 antigen and related molecules are involved in cell adhesion. Results presented below suggest that this isolated molecule can act as a substrate for axon outgrowth. To our knowledge, this is the first example of a cell surface protein in the CNS that has such an activity. Thus, the 8D9 antigen is critical in fasciculated outgrowth of axons not merely to hold axons together in a fascicle but to provide a pathway for the growth of subsequent axons. Since sequential outgrowth of axons along existing axons is important in many regions of the developing central nervous system (such as the developing optic nerve), the regulation of 8D9 antigen mediated interactions may be important in controlling these events.

The possibility that the 8D9 antigen functions in axon fasciculation and guidance suggests that its presence or absence may play a crucial role in the regeneration of damaged axonal pathways. The developmental expression of L1, NILE and Ng-CAM has been investigated in rat, mouse and chick (Nieke and Schachner, 1985; Stallcup et al., 1985; Thiery et al., 1985; Daniloff, et al., 1986). Although there are some species specific differences in the expression of these molecules, all are found in association with fascicles of axons during development; in mature peripheral nerves and the mature CNS, all are observed only on unmyelinated axons. It is possible that the absence of this molecule from mature myelinated CNS axon pathways could prevent the regeneration of axons to their targets. Strikingly, Nieke and Schachner (1985) find that L1 becomes strongly expressed on Schwann cells found in the distal stumps of transected sciatic nerves. Since peripheral nerves can support regeneration, it could be speculated that the reexpression of L1 reported by these authors represents the reestablishment of a permissive pathway for axon regeneration.

MATERIALS AND METHODS

MATERIALS AND ANIMALS.

Nitrocellulose was obtained from Schleicher and Schuell (Type BA 85). Laminin was obtained from Gibco; poly-l-ornithine (P-ORN) was obtained from Sigma and bacteriological petri plates were obtained from Falcon. Fertilized chicken eggs were obtained from Sachs and Sons Poultry Farms, Evans City, PA and incubated and staged as previously described (Lemmon and McLoon, 1986). Mice were outbred CD-1 obtained from Charles River, Wilmington, MA.

CELL CULTURE

Cell culture substrates were prepared by coating petri plates with nitrocellulose. This was done by dissolving 5 cm2 of nitrocellulose in 12 mls of methanol. Aliquots of 0.2 ml of this solution were rapidly spread over the surface of 60 mm petri plates and allowed to dry under a laminar flow hood. Test protein samples were applied in 1-5 Ml droplets containing 0.1-1.0 mg/ml, as specified. After approximately 1 minute, the droplets were removed by aspiration and the substrate plates were then blocked by washing twice with either DME/1% BSA or culture medium (DME with 10% horse serum for chick tecta; BME Earles with Hank's Salts and 10% horse serum for mouse cerebellum). Freshly dissociated tectal cells from embryonic day 10 (E10) chick embryos or cerebellar cells from postnatal day 6 (P6) mice were prepared as previously described (Lemmon and McLoon, 1986; Snitzer and Schachner, 1981) and added to substrate test plates in concentrations of 1 x 106/ml in 2 ml of culture medium. For some experiments, granule cells were purified from mouse cerebellum using Percoll (Pharmacia) gradients (Keilhauer et al., 1985). Mouse cerebellar explants were prepared by mechanical dissociation with fire polished Pasteur pipettes and cultured as described above.

QUANTIFICATION OF NEURITE LENGTH

The measurement techniques used were based on those developed by Chang et al., (1987). Culture dishes were examined with a Leitz Dilux inverted phase microscope equipped with a Dage-MTI-65 video camera which was interfaced to an IBM-XT computer equipped with the Bioquant Image Analysis System. Neurite length was measured as the distance between the center of the cell soma and the tip of its longest neurite. Neurites were only counted if a) the neurite emerged from a cell in isolation (not in a clump of cells), b) the neurite did not contact other cells or neurites, and c) and the neurite was longer than the diameter of the cell body. All cells with neurites meeting these criteria in an area of approximately 4.8 mm^2 were measured. The Mann-Whitney U test was used to determine whether different substrates produced significantly different amounts of neurite outgrowth. In addition, the percentage of single cells with neurites and the total number of cells in an area of 1.6 mm^2 were determined.

PURIFICATION OF 8D9 ANTIGEN AND N-CAM

These two cell adhesion molecules were affinity purified from embryonic chick brain using Affi-gel 10 (BioRad) coupled to either MAB 8D9 (Lemmon and McLoon, 1986) or MAB 224-1A6, which binds to N-CAM (Lemmon et al., 1982) as previously described except that membranes were solubilized in 1% sodium deoxycholate. Antigen was eluted with 0.1M diethylamine, pH 11.5 and immediately neutralized with solid Tris-HCl to use in the substrate test. To examine the purity of the 8D9 antigen and to compare it with other related molecules, 2-dimensional gels were prepared using isoelectric focusing (IEF) in the first dimension (using Servalyte 5-8 as ampholytes) and SDS PAGE for the second dimension as described by O'Farrell (1975).

IMMUNOCYTOCHEMISTRY

Mouse cerebellar astroglial cells were identified by staining with rabbit anti-GFAP antiserum (Dako); Chick tectal glial cells were identified by staining with MAB 3A7 (Lemmon, 1985). Rabbit anti-GFAP was detected with goat anti-rabbit IgG conjugated to HRP (Cappel Labs); HRP was detected with 4-chloro-1-napthol (Sigma). MAB 3A7 was detected with goat anti-mouse IgG conjugated with fluorescein (Cappel) and observed with a Leitz Ortholux phase-epifluorescence microscope.

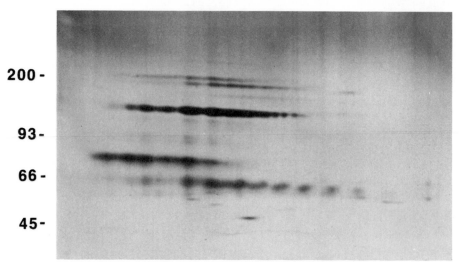

5.3   **7.0**

Figure 1. Two-dimensional gel analysis of 8D9 antigen. Affinity purified 8D9 antigen was first prepared by IEF in the 5-8 pH range and then run on a 5-15% gradient gel by SDS PAGE. The positions of molecular weight standards are indicated at the left and pH values are shown at the top. The proteins were visualized by silver staining.

RESULTS

IDENTIFICATION OF A NILE/L1/Ng-CAM LIKE MOLECULE

In 1983 we immunized mice with optic nerves from E14 chicks and obtained several different MABs that bound to axons of retinal ganglion cells. One of the MABs, designated 8D9 was found to bind cell surface proteins of 190,135,83 and 66 kilo-Daltons. The 8D9 antigen is similar in molecular weight and histological distribution to L1 in the mouse and Ng-CAM in the chick. We have demonstrated that polyclonal antisera prepared against mouse L1 (a generous gift of F. Rathjen and U. Rutishauser) cross-reacts with our affinity purified 8D9 antigen (Lemmon & McLoon, 1986). To further characterize the 8D9 antigen and more rigorously compare it to chick G4 (Rathjen et al., 1987), 2-dimensional gels using IEF and SDS PAGE were run. The resulting pattern shown in figure 1 is very similar to that of G4 (see Rathjen et al., 1987, figure 2A), a chicken molecule of which the N-terminal sequence is 50% homologous to that of mouse L1.

ASSAY FOR PROMOTION OF NEURITE OUTGROWTH

We wished to develop an assay system that would allow us to assess the ability of individual membrane-derived molecules to support attachment of neural cells. Since nitrocellulose is a convenient substrate for rapid noncovalent attachment of proteins, we coated sterile petri plates with nitrocellulose dissolved in methanol. Six substances were compared for their ability to promote neurite outgrowth. These included, laminin, P-ORN, N-CAM, 8D9 antigen, MAB 8D9 and MAB 224-1A6. To test the ability of nitrocellulose coated plates to attach substances that were known to act as cell attachment substrates, 1-5 Ml of P-ORN (0.1 mg/ml) and laminin (1mg/ml) were spotted on the coated plates with micropipettes and the plates' remaining binding capacity was blocked by washing with tissue culture medium containing 10% horse serum. Both substances were effective in attaching cells, but the morphology of the cells differed greatly. Laminin was very effective in supporting neurite outgrowth (figures 2B, 3B, 5C and 5D) in accordance with the findings of others (Rodgers et al., 1983); neurites on laminin were often fasciculated, although this was not the exclusive mode of outgrowth. Additionally, laminin promoted rapid spreading of astrocytes, oligodendrocytes, and fibroblast-like cells. In contrast, P-ORN was relatively poor in supporting neurite outgrowth (figures 3A, 5A and 5B). In preliminary experiments with P-ORN a concentration of 0.1 mg/ml was found to give optimum neurite outgrowth and was used for all subsequent experiments. Nitrocellulose coated only with serum showed relatively few fibroblast-like cells without attached neurons (figure 1D).

TEST OF MEMBRANE DERIVED MOLECULES

Two solubilized purified cell surface molecules that have been implicated in cell adhesion, the 8D9 antigen and N-CAM, were tested as substrates for cell attachment. The concentration of the 8D9 antigen spotted on the dishes was 0.11 mg/ml and that of N-CAM was 0.14 mg/ml. Since it was possible that the purified antigens contained traces of antibody from the affinity columns, antibodies were also tested as substrates. The results of experiments with dissociated chick tecta are shown in figure 2. Chick neurons attach rapidly (within 1-2 hours of plating) and extend long unfasciculated neurites on 8D9 antigen (figure 2A). The MAB 8D9 also attached neurons and induced some neurite outgrowth (figure 2C). In contrast,

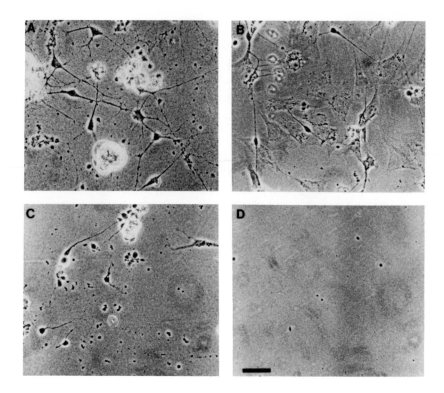

Figure 2. Chick tectal cells plated on various substrates. A) shows
several neurons with long neurites on 8D9 antigen. There are no flat
epithelial cells present. B) Cells plated on laminin. There are
numerous epithelial-like cells and a few neurons with long neurites.
C) Cells plated on MAB 8D9. There are some neurons with relatively
short neurites present. D) A region of the nitrocellulose coated
dish which had no substrate other than serum proteins. Neither
neurons nor flat cells find this to be an effective substrate for
attachment. Cells were in culture for 2 days. Scale bar = 20
microns.

N-CAM was unable to support significant cell attachment of chick
neural cells (data not shown). The ability of 8D9 antigen to support
unfasciculated neurite outgrowth was further explored with the use
of explant cultures of mouse cerebellum. As shown in figure 3,
micro-explants produced extremely thick neurite fascicles on laminin
or P-ORN but showed virtually no fasciculation when in contact with
the 8D9 antigen coated substrate. In cases where neurites exited the
explants above the surface of the plate, fasciculation was observed
up to the point of contact of the neurites with the 8D9 antigen

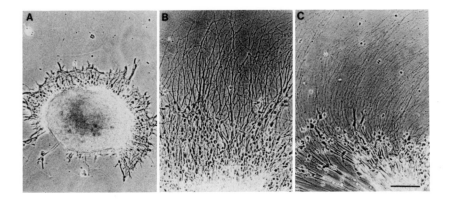

Figure 3.   A comparison of mouse cerebellar explant neurite outgrowth on differing substrates. All explants were prepared from P6 mice by mechanical dissociation and maintained in vitro for 3 days. Prior to plating explants, 1 Ml of various substrates were spotted on a nitrocellulose coated plate. A) Poly-ornithine resulted in relatively short, highly fasciculated neurites. B) Laminin produced long fasciculated neurites. C) 8D9 antigen produced long unfasciculated neurites. Scale bar = 100 microns.

coated substrate. Both MAB 8D9 and MAB 224-1A6 (which reacts with chick N-CAM) supported attachment of chick retinal cells but were unable to attach mouse cerebellar cells to the substrate (data not shown). These results suggest that the activity associated with 8D9 antigen is in fact due to interaction of mouse cells with the antigen and not trace amounts of antibody that were released from the affinity column.

QUANTIFICATION OF NEURITE OUTGROWTH ON DIFFERENT SUBSTRATES

Dissociated mouse cerebellar cells or purified cerebellar granule cells were plated on dishes containing non-overlapping sectors of P-ORN, laminin, N-CAM, 8D9 antigen, and an uncoated sector. After approximately 18 hours the cultures were fixed. Cell numbers and neurite lengths were then determined for each substrate as described in Materials and Methods. Figure 4 shows the data from one experiment. The 8D9 antigen produced the longest neurites; some exceeding 200 microns in length after 18 hours in culture. Laminin also produced cells with long neurites, but the longest was less than 150 microns. P-ORN and N-CAM were both much less effective at producing neurite outgrowth, although they were better than uncoated

122

Table 1. Neurite growth on different substrates

SUBSTRATE	SINGLE CELLS WITH NEURITES n	NEURITE LENGTH,* μm	% SINGLE CELLS WITH NEURITES¶	TOTAL CELLS (SINGLE & CLUMPED)¶ n
EXPERIMENT 1, PURIFIED GRANULE CELLS, 18 HR IN CULTURE				
8D9 antigen	59	56+4	56	64
P-ORN	51	31+2	19	213
Laminin	57	49+4	51	312
N-CAM	11	23+3	31	23
Uncoated	-	-	0	4
EXPERIMENT 2, CEREBELLAR CELLS, 18 HR IN CULTURE				
8D9 antigen	53	66+6	66	258
P-ORN	52	35+2	29	551
Laminin	56	63+7	47	334
N-CAM	14	44+9	30	28
Uncoated	-	-	0	6

* Mean + SEM
¶ In 1.6 mm²

tissue culture plastic. As summarizd in table 1, 8D9 antigen produced the longest neurites although laminin was also an effective substrate for neurite outgrowth. The 8D9 antigen was significantly better than P-ORN, N-CAM and tissue culture plastic coated only with nitrocellulose at the 0.01 level or better. Similarly, 8D9 antigen produced the highest percentage of cells with neurites. Both P-ORN and laminin were better as cell attachment factors than 8D9 antigen. N-CAM was a relatively ineffective substrate for either cell attachment or neurite outgrowth, being only slightly better than uncoated plates. Dilution of the 8D9 antigen indicated that below a concentration of 0.01 mg/ml it was ineffective as a substrate for cell attachement or neurite outgrowth.

CELL CLASS SPECIFICITY

To determine if the 8D9 antigen supported attachment of one class of neural cells in preference to others, dissociated mouse cerebellar cells were plated on 8D9 antigen. After three days, the

Figure 4. Distribution of neurite lengths of mouse cerebellar granule cells grown on different substrates. The percentage of neurons with neurites (vertical axis) longer than a given length (horizontal axis). The distributions are shown for cells on 8D9 antigen, P-ORN, laminin, N-CAM and a control region of the dish that received no substrate other than the BSA blocking solution.

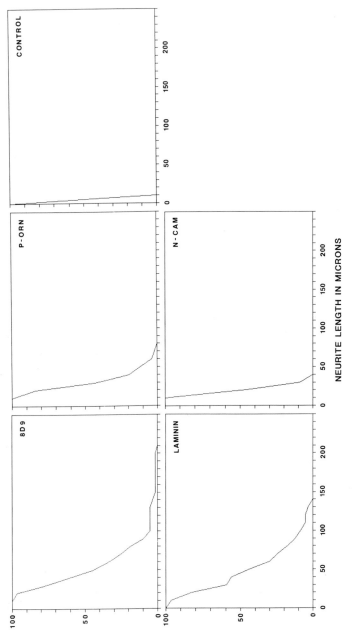

% OF CELLS WITH NEURITES GREATER THAN LENGTH ON X-AXIS

NEURITE LENGTH IN MICRONS

cells were stained with anti—GFAP to assess the relative numbers of astrocytes. For comparison, P—ORN and laminin coated substrates that had been seeded with cells from the same dissociation were also stained for GFAP. Both P—ORN and laminin were effective in attaching neurons and glia, although laminin was a greatly preferred substrate for both cell types (figures 5A—D). As shown in figures 5E & 5F, 8D9 antigen coated substrate was almost free of astrocytes. In fact, most astrocytes that were detectable on the 8D9 antigen coated substrate were observed in association with clumps of neurons, suggesting that their attachment was via the neurons and not as a result of interaction with 8D9 antigen. Parallel studies using chick cells and antibody 3A7 (Lemmon, 1985) for the detection of chick glia gave similar results (data not shown).

DISCUSSION

An important feature of this work was the development of a novel means of attaching purified proteins to cell culture dishes to facilitate identification of novel cell attachment factors. We hypothesised that methanol-solubilized nitrocellulose might greatly increase the protein binding capacity of tissue culture plastic. Unpublished tests in our laboratories indicate that nitrocellulose coated plastic bound approximately 15 times as much ^{125}I labeled protein as untreated tissue culture plastic. In addition, the retention time of protein on nitrocellulose coated plastic was several times longer than that on untreated plastic.

Figure 5. Cell type specificity of cerebellar cell attachment and growth on different substrates. Cells were maintained in vitro for 3 days and then stained for glial fibrillary acidic protein (GFAP) to detect astrocytes. HRP and 4-chloro-1-napthol were used to visualized the location of the anti—GFAP. A, C, and E are phase photographs to demonstrate the location of neurons and other cells. B, D, and F are transmitted light photographs to demonstrate the GFAP containing cells. A and B) Poly-ornithine was an effective substrate for both neurons and GFAP positive cells. C and D) Laminin was a very effective substrate for neurons and GFAP positive cells. E and F) 8D9 antigen was a highly effective substrate for neuronal attachment and resulted in the production of numerous neurites. GFAP positive cells were present in much lower numbers than on laminin and appeared to be present in aggregates with neurons. Scale bar = 100 microns.

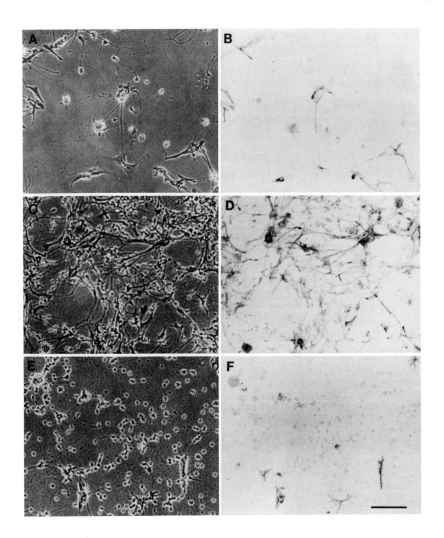

It also is possible that nitrocellulose binds the membrane proteins in some optimized orientation. In any case, we have never been able to obtain cell attachment or neurite outgrowth on untreated tissue culture plastic that was incubated with 8D9 antigen or N-CAM.

The use of methanol-solubilized nitrocellulose has a number of advantages over drying proteins on tissue culture plastic, attaching them via reactive cross-linkers or growing cells on sheets of nitrocellulose. These include the fact that the resulting nitrocellulose coating is clear and permits excellent optical observation of cells. Plate preparation is very easy, taking less than 10 minutes from dissolving the nitrocellulose until it is dry on the dishes. Proteins bind to the nitrocellulose in less than a minute and the drops do not have to be evaporated to insure adequate attachment. This method allows one to spot many samples on a single plate. Therefore, it is possible to assay numerous fractions, such as those from a molecular sieve column, for cell attachment activity. Nitrocellulose does not appear to be toxic to neural cells and is a very poor substrate for neural cell attachment when coated only with bovine serum albumin or serum. This gives low levels of background attachment and makes identification of true cell attachment molecules relatively easy. Nitrocellulose may not be advantageous for molecules such as laminin, which bind to tissue culture plastic when given enough time.

The findings presented here identify an activity associated with the isolated 8D9 antigen. When attached to nitrocellulose coated plates, 8D9 antigen supports the attachment of both embryonic chick tectal and postnatal mouse cerebellar neurons and the rapid extension of neurites. Quantitative measurement of neurite length demonstrated that 8D9 antigen promoted neurite outgrowth comparable to or better than laminin. The 8D9 antigen was not as effective at promoting cell attachment as laminin or P—ORN, perhaps indicating that the 8D9 receptor is predominantly localized to axons, as is the 8D9 antigen (Lemmon and McLoon, 1986). Neurites from both sources appear to be unfasciculated when they grow on 8D9 antigen. These results are in agreement with the proposed involvement of 8D9 antigen related molecules NILE and Ng–CAM in fasciculation. These results seem in conflict with those of Stallcup and Beasley (1985) who have compared the effects of monovalent antibodies directed against NILE and N-CAM. These workers demonstrated that antibodies directed against NILE inhibited fasciculation in cultures of embryonic rat cerebellum but could not do so in cultures prepared from postnatal day 5 rats. They were, however, able to demonstrate inhibition of fasciculation in P5 rat cerebellar cultures with antibodies directed against N-CAM. We believe that our results may differ for several reasons. It is possible that the receptor for the 8D9 antigen is maintained at high levels in both early and later development while the 8D9 antigen is not; in this case, our assay which provides exogenous, purified 8D9 antigen would reveal unfasciculated outgrowth at both stages. Alternatively, immobilized 8D9 antigen may provide a more effective substrate than mobile N-CAM and 8D9 antigen on other nearby axons. The 8D9 antigen coated plates also present axons with a single choice of adhesive substrate without other possible adhesive (or repulsive) cell surface molecules. Although these differences may significantly affect the types of interactions that are possible between neurons and 8D9 antigen, this assay system does provide a straight-forward method for analyzing the adhesive interactions of 8D9 antigen with cells and other molecules.

A remarkable finding of these studies is that both chick and mouse neurons can attach and extend neurites on chick derived 8D9 antigen. This suggests that the site on the 8D9 antigen responsible for neurite extension has been highly conserved over evolution. Interestingly, poor immunological cross reactivity of the 8D9 related molecule G4 with mouse L1 antigen and n-terminal sequence homology studies (Rathjen et al., 1987) indicate only partial homology between the chick and mouse molecules. This suggests that the portion of 8D9 antigen that is involved in cell-cell binding has been selectively conserved. While neurons are able to rapidly attach to 8D9 antigen and send out neurites, no preferential attachment of glia was seen with either test culture system, although both contained many nonneuronal cells that could be cultured on laminin or P-ORN bound to nitrocellulose. Although our findings clearly support the role of 8D9 antigen in neuron-neuron and not neuron-glia adhesion, in agreement with others (Rathjen et al., 1987; Keilhauer et al., 1985; Chang et al., 1987), it is possible that glial adhesion is weaker or that in the course of isolation of 8D9 antigen, a glial binding site has been destroyed or obscured in binding to nitrocellulose.

To date two classes of cell surface molecules have been implicated in axon fasciculation: adhesion molecules related to the

8D9 antigen and those related to N-CAM. Fab fragments of antibodies directed against these molecules were demonstrated to interfere with fasciculated axon outgrowth from explants of neural tissue (Stallcup and Beasly, 1985; Hoffman et al., 1986). Our findings indicate that the role of 8D9 antigen on a given axon is to promote the extension of other neurites along this axon which in turn produces fasciculated axons. We were unable to demonstrate a similar activity with immuno-purified N-CAM antigen. This may simply indicate that N-CAM is difficult to purify in an active form or that it was unable to bind to nitrocellulose coated plates in an active orientation. Alternatively, these findings could indicate that N-CAM does not play a major role in fasciculated axon outgrowth at the stages we examined. Our results concerning the effectiveness of 8D9 and the ineffectiveness of N-CAM as substrates for neurite outgrowth are also in agreement with the results of Rathjen and associates (Rathjen et al., 1987; Chang et al., 1987).

Studies of the 8D9 related adhesion molecule, Ng-CAM, indicate that binding is heterophilic (Hoffman et al., 1986), although, the receptor for it has not been isolated. Data presented here indicates that the 8D9 receptor is present on both neurites and growth cones, because both structures attach readily to 8D9 antigen coated dishes. Since the isolated molecule retains binding activity it should be possible to use 8D9 antigen as an affinity ligand for identification of the 8D9 receptor.

The possibility that the 8D9 antigen functions in axon fasciculation and guidance suggests that its presence or absence may play a crucial role in the regeneration of damaged axonal pathways. The developmental expression of L1, NILE and Ng-CAM has been investigated in rat, mouse and chick (Nieke and Schachner, 1985; Stalcup et al., 1985; Thiery et al., 1985; Daniloff et al., 1986). Although there are some species with specific differences in the expression of these molecules, all are found in association with fascicles of axons during development; in mature peripheral nerves and mature CNS, all are observed only on unmyelinated axons. It is possible that the absence of this molecule from mature myelinated CNS axon pathways could prevent the regeneration of axons to their targets. Strikingly, Schachner and co-workers find that L1 becomes strongly expressed on Schwann cells found in the distal stumps of transected sciatic nerves (Nieke and Schachner, 1985). Since periphial nerves can support regeneration, it could be speculated that the reexpression of L1 represents the reestablishment of a permissive pathway for axon regeneration.

Based on the studies presented here, 8D9 antigen appears to be a good choice for a substrate in regeneration experiments. Is it possible that purified 8D9 antigen could promote axon regeneration in peripheral nerves where suitable nerve grafts are unavailable? Could the purified 8D9 antigen be used as a substrate for regeneration of CNS axon pathways? In vivo experiments will be required to test these hypotheses.

ACKNOWLEDGEMENTS

We would like to thank J. Hailey, L. Lowenadler, and D. Memon for their excellent technical assistance. This work was supported by a Basil O'Conner Grant (#5-518) to C.L. and a Basic Research Grant (#1-979) to V.L., both from the March of Dimes, and a grant from the National Eye Institute (EY-5285) to V.L. Some of the results described in this paper have been published elsewere (Lagenaur & Lemmon, 1987).

REFERENCES

Bock, E., C. Richter-Landsberg, A. Faissner, & M. Schachner. 1985. Demonstration of immunochemical identity between the nerve growth factor-inducible large external (NILE) glycoprotein and the cell adhesion molecule L1. EMBO J. 4:2765-2768.

Bozyczko, D. & A.F. Horwitz. 1986. The participation of a putative cell surface receptor for laminin and fibronectin in peripheral neurite extension. J. Neuroscience 6:1241-1251.

Chang, S., F.G. Rathjen, and J.A. Raper. 1987. Extension of neurites on axons is impaired by antibodies against specific neural surface glycoproteins. J. Cell Biology. 104:355-362.

Cole, G.J., D. Schubert, and L. Glaser. 1985. Cell-substratum adhesion in chick neural retina depends upon protein-heparan sulfate interactions. J. Cell Biol. 100:1192-1199

Daniloff, J.K., C.-M. Chuong, G. Levi, & G.M. Edelman. 1986. Differential distribution of cell adhesion molecules during histogenesis of the chick nervous system. J. Neuroscience. 6:739-758.

Davis, G.E., M. Manthorpe, E. Engvall, & S. Varon. 1985. Isolation and characterization of rat Schwannoma neurite-promoting factor: evidence that the factor contains laminin. J. Neuroscience 5: 2662-2671.

Edelman, G.M. 1985. Cell adhesion and the molecular processes of morphogenesis. Annu. Rev. Biochem. 54:135-169.

Friedlander, D.R., M. Grumet, & G.M. Edelman. 1986. Nerve growth factor enhances expression of neuron-glia cell adhesion molecule in PC12 cells. J. Cell Biol. 102:413-419.

Grumet, M., S. Hoffman, C.-M. Chuong, and G.M. Edelman. 1984. Polypeptide components and binding functions of neuron-glia cell adhesion molecules. PNAS 81:7989-7993.

Grumet, M., U. Rutishauser, and G. Edelman. 1983. Neuron-glia adhesion is inhibited by antibodies to neural determinants. Science 222:60-62.

Hatta, K. and M. Takeichi. 1986. Expression of N-cadherin adhesion molecules associated with early morphogenetic events in chick development. Nature 320:447-449.

Hoffman, S., D.R. Friedlander, C.M. Choung, M. Grumet, & G.M. Edelman. 1986. Differential contributions of Ng-CAM and N-CAM to cell adhesion in different neural regions. J. Cell Biol. 103: 145-158

Keilhauer, G., A. Faissner, & M. Schachner. 1985. Differential inhibition of neurone-neurone, neurone-astrocyte and astrocyte-astrocyte adhesion by L1, L2, and N-CAM antibodies. Nature, 316:728-730.

Kruse, J., G. Keilhauer, A. Faissner, R. Timpl, & M. Schachner. 1985. The J1 glycoprotein - a novel nervous system cell adhesion molecule of the L2/HNK-1 family. Nature 316:146-148.

Lagneaur, C., and V. Lemmon. Nov., 1987. An L1-like molecule, the 8D9 antigen, is a potent substrate for neurite extension. Proc. Natl. Acad. Sci. U.S.A. In Press.

Lander, A.D., D.K. Fujii, and L.F. Reichardt. 1985. Laminin is associated with the "neurite outgrowth-promoting factors" found in conditioned media. Proc. Nat. Acad. Sci. U.S.A. 82:2183-2187.

Lemmon, V. 1985. Monoclonal antibodies specific for glia in the chick nervous system. Dev. Brain Res. 23:111-120

Lemmon, V., Staros, E.B., Perry, H.E., and Gottlieb, D.I. 1981. A monoclonal antibody which binds to the surface of chick brain cells and myotubes: Cell selectivity and properties of the antigen. Dev. Brain Res. 255:349-360.

Lemmon, V., & S.C. McLoon. 1986. The appearance of an L1-like molecule in the chick primary visual pathway. J. Neurosci. in press

Letourneau, P.C. 1975. Cell-to-substratum adhesion and guidance of axonal elongation. Dev. Biol. 44:92-101.

Lindner, F.G. Rathjen and M. Schachner. 1983. L1 mono- and polyclonal antibodies modify cell migration in early postnatal mouse cerebellum. Nature 305:427-430

McGuire, J.C., L.A. Greene, and A.V. Furano. 1978. NGF stimulates incorporation of fucose or glucosamine into an external glycoprotein in cultured rat PC12 pheochromocytoma cells. Cell 115:357-365.

Nieke, J. and M. Schachner. 1985. Expression of the neural cell adhesion molecules L1 and N-CAM and their common carbohydrate epitope L2/HNK-1 during development and after transection of the mouse sciatic nerve. Differentiation 30:141-151.

Nobel, M., M. Albrechtsen, C. Moller, J. Lyles, E. Bock, C. Goridis, M. Watanabe, & U. Rutishauser. 1985. Glial cells express N-CAM/D2-CAM-like polypeptides in vitro. Nature 316:725-728.

O'Farrel, P.H. 1975. High resolution two-dimensional electrophoresis of proteins. J. Biol. Chem. 250:4007-4021.

Rathjen, F.G. and M. Schachner. 1984. Immunocytological and biochemical characterization of a new neuronal cell surface component (L1 antigen) which is involved in cell adhesion. EMBO Journal 3: 1-10.

Rathjen, F.G., Wolff, J.M., Frank, R., Bonhoeffer, F., Rutishauser, U., and Schoeffski, A. 1987. Membrane Glycoproteins involved in neurite fasciculation. J. Cell Biol. 104:343-353.

Rodgers, S.L., LeTourneau, P.C., Palm, S.J., McCarthy, J., and Furcht, L.T. 1983 Neurite extension by peripheral and central nervous system neurons in response to substratum-bound fibronectin and laminin. Dev. Biol.98:212-220.

Smalheiser, N.R., S.M. Crain, and L.M. Reid. 1984. Laminin as a substrate for retinal axons in vitro. Dev. Brain Res. 12:136-140.

Snitzer, J. and Schachner, M. 1981. Expression of Thy-1, H-2 and NS-4 cell surface antigens and tetanus toxin receptors in the developing and adult mouse cerebellum. J. Neuroimmunol. 1:429-456.

Stallcup, W.B., L.S. Arner, and J. Levine. 1983. An antiserum against the PC12 cell line defines cell surface antigens specific for neurons and Schwann cells. J. Neuroscience 3:53-68.

Stallcup, W.B., & L.L. Beasly. 1985. Involvement of the nerve growth factor-inducible large external glycoprotein (NILE) in neurite fasciculation in primary cultures of rat brain PNAS 82:1276-1280.

Stallcup, W.B., L.L. Beasley, and J.M. Levine. 1985. Antibody against nerve growth factor-inducible large external (NILE) glycoprotein labels nerve fiber tracts in the developing rat nervous system. J. Neuroscience 5:1090-1101.

Thiery, J.-P., A. Delouvee, M. Grumet and G.M. Edelman. 1985. Initial appearance and regional distribution of the neuron-glia cell adhesion molecule in the chick embryo. J. Cell Biol. 100:442-456.

Cognin and Retinal Cell Differentiation
Robert E. Hausman

The initial work on what is now known as retina cognin was carried out in Aron Moscona's laboratory at the University of Chicago. In fact, the idea of cell-type specific markers was one in which he was an early participant, from the early 1950's onward. Moscona's term for these macromolecules was "ligands" (Moscona, 1968). Early observations suggested that there was a conditioned medium activity from embryonic chick neural retina cells which was able to enhance the normal aggregation in vitro of these cells (Moscona, 1963). Although retina cells without the added material would aggregate in gyratory rotation culture, the size of the aggregates was significantly greater (usually several-fold) in the presence of the added material. In all cases, the cells within the aggregates reorganized into histotypic patterns (Moscona, 1956; 1957; 1961; 1962). This activity was tissue specific in that the culture supernatant from retina cells had no similar effect on the aggregation of cells from several other tissues. Lilien, working in Moscona's laboratory, followed up on these observations and demonstrated that this specific cell aggregation-enhancing activity was independent of serum in the culture (Lilien and Moscona, 1967; Lilien, 1968). Garber, Hausman and others working in association with Moscona, showed that cultures of cells from other tissues produce conditioned media with their own specific cell aggregation-enhancing properties. This is true even of cells from other regions of the embryonic chick CNS, such as optic tectum and cerebrum (Garber and Moscona, 1972; Hausman et al., 1976).

Few of these other cell aggregation-enhancing activities (possible cognins) have been further investigated. However, McClay and Hausman began the purification of the aggregation-enhancing activity from neural retinal cell culture medium. We worked out the rough size and characteristics of the macromolecule and showed that it was synthesized by the cells as part of a small class of proteins whose synthesis can be blocked by proflavine, a diamino acridine (Hausman and Moscona, 1973; McClay and Moscona, 1974).

Retina cognin is a 50 kd glycoprotein whose amino acid and sugar compositions are known. However, the amino acid sequence remains unknown largely because of the difficulty of obtaining enough protein. The amino acids present are not noteworthy, and the sugars are unexceptional. Cognin contains about 1/3 each of glucosamine, mannose, galactose with just enough sialic acid to be explained by chain terminal residues. This ubiquity of the sugar residues is a likely explanation for the difficulty in obtaining monoclonal antibodies to cognin. In contrast to the neural cell adhesion molecule (N-CAM) (Hoffman et al., 1982), the size and biophysical properties of cognin suggest no extraordinary shape (Hausman and Moscona, 1975; Moscona and Hausman, 1977).

A final step in the initial characterization of cognin was determining if the protein was purely a secretory product of the cells in culture or was it directly associated with the cell membrane. It was not known at the time, that membrane vesicles and other molecular complexes (Schmitt et al., 1985) are often shed into the culture medium in vitro. We demonstrated that cognin could not be removed from the cell surface by agents which typically remove exterior peripheral proteins (high salt, low pH) but could be removed by disrupting the lipid bilayer (with organic solvents or detergents) (Hausman and Moscona, 1976b). We went on to demonstrate that cognin could be prepared directly from a membrane preparation obtained from embryonic chick retinas (Hausman and Moscona, 1976a). At the time, this suggested to us that cognin is an intrinsic protein to the membrane, but the recent discovery of phosphoinositol and other types of protein to lipid linkages (Marchase, 1987) raises other possibilities. All the available evidence (see immunocytochemisty below) is consistent with cognin being membrane associated and not normally occurring in the cytosol or in the extracellular matrix.

RETINA COGNIN ACTIVITY IN VITRO.

Cognin enhances the aggregation of embryonic chick neural cells from a broad spectrum of early and middle embryonic ages (see below). It has no similar effect on embryonic chick cells isolated from other tissues tested. This is what is meant by the term "tissue-specificity" when applied to cognin. The in vitro aggregation-enhancing effect of cognin is most evident after prior treatment of the cell membrane with trypsin, perhaps because the cognin already present on the retina cell surface is removed (Moscona and Hausman, 1977; Hausman and Moscona, 1979). The large aggregates generated by cognin are qualitatively different from the clumps of cells generated by lectins or other non-endogenous cell-cell ligands. Cells within the cognin enhanced aggregates are able to move about and reorganize into characteristic histotypic retina tissue. Cells within the clumps generated by other ligands either remain fixed in place or the ligand must first be metabolically removed from the surface before histogenesis can subsequently occur (Moscona and Hausman, 1977).

Inhibition of general protein synthesis blocks retina cell aggregation and the aggregation-enhancing effects of cognin, perhaps because other macromolecules necessary for its activity must be synthesized (see below) (Moscona, 1962). However, blocking cognin synthesis 'preferentially' with proflavine (which blocks less than 20% of all protein synthesis) blocks normal aggregation in a manner which can be rescued by the addition of exogenous cognin. This result strongly suggests that cognin is necessary for normal cell aggregation following dissociation of the cells from tissue with trypsin (Hausman and Moscona, 1973). The inhibition of biosynthesis caused by proflavin does not extend to the molecules associated with cognin.

As might be expected, polyclonal antibodies to cognin block the aggregation of neural retina cells if they are added to cells which have cognin on their surfaces (Hausman and Moscona, 1979). Such cells can be generated by allowing freshly trypsinized retina cells to replace membrane proteins (including cognin) under conditions where the cells cannot aggregate (absence of Ca^{2+} or high speeds of gyratory rotation). Cognin antibodies also block the reaggregation of membrane vesicles which are prepared directly from neural retina tissue without trypsin treatment. This effect of cognin antibodies is also specific to neural retina cell membranes. Cognin antibodies have no effect on the aggregation of membrane vesicles prepared from other neural tissues (Troccoli and Hausman, 1985). Thus, the results appear to be quite different from the multiple effects of antibodies to cell adhesion molecules such as N-CAM which affect the cells of many different embryonic tissues, often in different ways (Rutishauser, 1984).

The name retina cognin was coined to denote the specific recognition properties of the molecule. We felt strongly that the significance of this type of molecule was not that it caused cell-cell adhesion, but that the cell-cell contact engendered allowed some significant differentiation event (with biochemical or physiological consequences) to occur in the cells in contact (Hausman and Moscona, 1976; Moscona and Hausman, 1977). Clearly, the commonly used in vitro assays quantitated only adhesion. However, for cognin the adhesions so generated allowed subsequent differentiation in vitro. Just as clearly, we do not yet know the nature of the significant in vitro differentiation event (or events) (Hausman et al., 1987).

THE DEVELOPMENTAL-DEPENDENCE OF COGNIN'S ACTIVITY IN VITRO.

It has long been known that the ability of embryonic cells in general, and retina cells in particular, to reassociate in vitro declines with embryonic age. The ability of retina cells to form histotypic aggregates declines as they are removed from older and older embryos (Moscona, 1961; 1962). Several lines of evidence have implicated changes in cognin function in this decline. Chick retina cells put into culture at 9-10 days of embryogenesis cease conditioning the culture medium after three days in culture; cells put into culture after 12 days of embryogenesis fail to condition the culture medium (Hausman and Moscona, 1973; 1979). These experiments suggested that the ability to produce cognin declined with embryogenesis. Additionally, retina cells isolated from embryos older than 12 days show an age-dependent loss of response to added cognin. Cell aggregation may be enhanced by the added cognin, but the sizes of the enhanced aggregates are smaller and many aggregates are of normal size (Hausman and Moscona, 1979). The cells acted as if fewer and fewer of them were able to respond to added cognin. Working in Moscona's laboratory, Ben-Shaul and

Hausman looked at embryonic retina cells from different ages, which had been allowed to replace the cognin on their surfaces. We found that fewer and fewer retina cells exhibited cognin with increasing embryonic age, a result consistent with the findings cited above (Ben-Shaul et al., 1979; 1980). At that time, we could not distinguish between the actual loss of cognin from most retina cell surfaces and a loss in the ability of older embryonic cells to recover from trypsin. However, the observation that cells maintained in culture for more than three days also lost the ability to produce cognin suggested that the loss might be a normal consequence of development, not a trypsin recovery artifact. Recent work which investigated the presence of cognin on membrane vesicles has confirmed that there is a natural decline in the amount of cognin on retina cell surfaces with increasing embryonic development. Furthermore, this decline in cognin on the membrane matches the declining ability of such vesicles to aggregate in vitro (Troccoli and Hausman, 1985).

THE BASIS OF COGNIN'S IN VITRO DEVELOPMENTAL-DEPENDENCE.

The explanation for the developmental dependence of cognin's effect is not yet clear and might involve the nature of the macromolecule or molecules to which it binds. Unlike N-CAM, cognin appears to bind one neural retina cell membrane to another via a heterophilic interaction. Two lines of evidence suggest that cognin does not bind to itself under in vitro conditions. Both affinity chromatography of retina membrane proteins on a cognin column and photoaffinity crosslinking to immunologically detectable cognin suggest that it binds other macromolecules. Cognin appears to bind to a protein or proteins of the size of 64 kd on polyacrylamide gel electrophoresis (Troccoli and Hausman, 1988). This is quite different from the cell adhesion activity of N-CAM which binds itself homophilically (Rutishauser et al., 1982).

Understanding the nature of the cognin-to-64 kd protein interaction and how it might affect the developmental dependence of cognin's aggregation-enhancing activity will require further investigation. Specifically, progress in this area will require the development of antibodies to the putative binding protein. This will allow us to compare the location of the 64 kd protein with the known changes in the distribution of cognin in the retina during

138

development (see below).

LOCATION OF COGNIN IN VIVO.

Cognin appears during normal development largely in the embryonic chick retina. Searches of the early avian embryo for cognin immunogenicity have failed to detect cognin in other regions of the embryo. Cognin is evident in the eye by embryonic day 3-4 and remains in the chick retina throughout embryonic development until after hatching. It is lost about 2-3 weeks after hatching in the chick. We have not investigated cognin in other embryonic retinas. While antisera cross react, immunological staining has not been satisfying. This might have to do with the sensitivity of the cognin molecule to fixatives (Dobi et al., 1986).

Cognin appears sequentially in two developmentally-dependent patterns in the embryonic chick retina. Cognin is found on apparently all cells up to about 10 days of embryonic development. At this time it begins to be lost from the surfaces of cells in the outer retina. First, it is lost from the rod and cone outer segments (though it is not evident it ever was there, the outer segments may just be more apparent at this stage) then from the photoreceptor cell bodies, from their processes and then from the cell bodies in the inner nuclear layer (Dobi et al., 1986). This pattern resembles a gradient of cognin expression on cell surfaces. The interesting question is if this pattern is itself the result of some gradient of control substance in the tissue.

To investigate this, we asked if the characteristic loss of cognin from retina cells could occur when the cells were removed from the tissue and allowed to continue their differentiation in vitro. The answer is that the characteristic loss of cognin from most of the neurons still occurs and it is retained on about 20% of the cells, a percentage similar to that within the ganglion cell layer (GCL) of the intact retina. Thus, it appears that there is no requirement for a gradient within the tissue to explain the apparent loss of cognin; instead the cells appear to be internally programmed by 9 days of embryonic development to lose or retain cognin on their surfaces (Hausman et al., 1986). This lack of position-dependence in the later differentiation steps of retinal neurons is consistent with the findings of Adler's group and others on photoreceptor differentiation in vitro (Adler et al., 1984).

POSSIBLE BASES FOR THE DEVELOPMENTALLY-DEPENDENT DISTRIBUTION OF COGNIN.

As a first step, we asked if there was any correlation between the presence of cognin on the membrane of retina cells and any specific type or direction of cell differentiation. Second, we asked if the location of cognin in the retina was similar to that of any other cellular differentiation marker. We had earlier noted a membrane-dependent association between cognin and the retinal alpha bungarotoxin receptor. When vesicles were prepared from retina tissue and fractionated on a alpha bungarotoxin affinity column, those vesicles which contained the alpha bungarotoxin receptor contained a disproportionate amount of cognin. As the age of the retina from which the membrane vesicles were prepared was increased, the correlation became more significant (Hausman et al., 1985). These findings seemed to indicate that as the alpha bungarotoxin receptor partitioned into a smaller and smaller fraction of the total cellular membrane in the retina, cognin partitioned into that same fraction of membrane. Just what this result meant for the physiological acetylcholine receptor was temporarily clouded by a dispute over the significance of alpha bungarotoxin binding in the CNS and in the retina in particular. Was alpha bungarotoxin binding to the functional acetylcholine receptor in the retina or not? With that dispute settled in favor of alpha bungarotoxin identifying the functional acetylcholine receptor, our results clearly suggested a spatial correlation between cognin and the acetylcholine receptor (Hausman et al., 1985).

The histological location of the acetylcholine receptor in the embryonic chick retina has been investigated by Nirenberg and his colleagues (Vogel and Nirenberg, 1976; Vogel et al., 1977). Their results indicated that the acetylcholine receptor occurred widely in the retina during early development and became concentrated in the inner plexiform layer (IPL) during the latter half of embryogenesis (Figure 1). Subsequent work has established that the acetylcholine receptor gradually becomes localized in specific lamina of the IPL where it occurs on the synaptic and extra synaptic membranes of the amacrine and ganglion cells (Puro, 1985) (Figure 2). These strata of the retina are one place where cognin localizes after 11 days of development. However, by comparing the tissue locations of alpha bungarotoxin binding and cognin antibody reaction, it was clear that

140

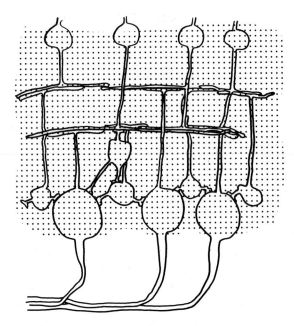

Figure 1. Schematic drawing of the cells bordering the inner plexiform layer (IPL) of the late embryonic chick neural retina. The distribution of alpha bungarotoxin binding over the IPL before embryonic day 14 is shown by dots (adapted from Puro, 1985). Shown at the top are some amacrine cells in the inner nuclear layer, simplified neurite strata and an inner plexiform cell in the IPL and the displaced amacrine cells in the ganglion cell layer (GCL). The axons of the large ganglion cells exiting at the bottom form the nerve fiber layer.

cognin was not concentrated in cholinergic synapses (Dobi et al., 1986). Instead it was found across the cell body and neurites of cells which exhibited alpha bungarotoxin binding (Figure 3). This suggested to us that, while there was a histological correlation between cognin and the actylcholine receptor, it did not extend to the level of the synapse. This suggested that cognin was not playing a direct role in nicotinic cholinergic synapse formation.

The correlations between cognin and alpha bungarotoxin binding as well as the disappearance of cognin from the adult avian retina suggested to us that cognin might play a role in the differentiation of retina ganglion cell layer (GCL) neurons.

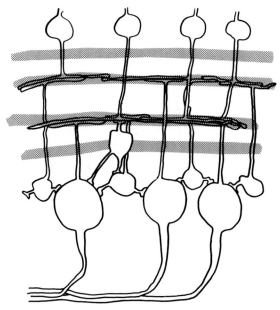

Figure 2. Schematic drawing of the cells bordering the inner plexiform layer (IPL) of the late embryonic chick neural retina. The distribution of alpha bungarotoxin binding in the IPL between embryonic day 14 and hatching at embryonic day 21 is shown (adapted from Puro, 1985). Cells are as described in the legend to Figure 1.

THE DIFFERENTIATION OF RETINAL NEURONS

The mechanisms of early cell differentiation in CNS tissues, including the retina, are not well understood (Meller, 1984). The steps by which the tissue-spanning ventricular cells (which appear homogeneous) become determined and eventually differentiate is one of the topics which are being actively investigated (see Raymond, this volume). A wide variety of approaches including classic thymidine birthdate analysis (Fujita and Horii, 1963; Kahn, 1974), have suggested that following a period of rapid cell division, there is a period of cellular commitment, migration to a characteristic position within the tissue and overt differentiation (Meller, 1984). However, the actual kinetics of this putative sequence of cell differentiation events have never been worked out. Nonetheless, there has been a widely-shared belief that migration and differentiation are correlated.

These ideas led Sidman to suggest the idea of "determination of address" (Sidman, 1975). Just what this idea of cell migration (perhaps along radial glia) and differentiation implied and to which

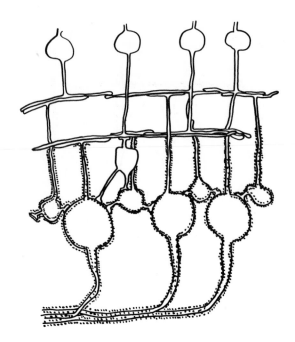

Figure 3. Schematic drawing of the cells bordering the inner plexiform layer (IPL) of the late embryonic chick neural retina. The distribution of cognin immunoreactivity in the retina between embryonic day 14 and about two weeks posthatching is shown. Cells are as described in the legend to Figure 1. The surfaces of displaced amacrine and ganglion cells are stained. The staining of ganglion cell axons is lost as they exit the eye in the optic nerve (Dobi et al., 1986).

CNS organs it applied has varied with the passage of time and our increasing understanding of the complexities of neural development. In the present context, we are interested in what it might mean for the possible function of cognin in the retina.

DETERMINATION OF CELL FATE

This idea of "address" touches on a question which is larger than the retina as a tissue and larger than the CNS as a group of tissues. That question is early, invarient determination versus flexibility of developmental pathways. For the CNS, this question has been brought to our attention over the past few years by Marcus Jacobson and others who have looked at initial embryonic induction of the nervous system (Jacobson, 1984). There is clearly some neural property (tendency) apparent early in the embryonic ectoderm. Perhaps it is this initial property which yields the ventricular cell histology which is characteristic of all early embryonic CNS tissue. However, the determination of cell-type specificity within each tissue of the CNS appears to be much more flexible and might occur much later in development. This might even be the case for the decision to become a neuron or a glial cell (see Linser, this volume).

Of all retinal neurons, the most extensive work has been on the two types of photoreceptors, rods and cones. When removed from the tissue, cells lose their characteristic morphology. Cones are able to regain their characteristic shape in culture and go on to exhibit many aspects of normal differentiation (Adler et al., 1984). Thus, cones clearly are able to differentiate en the absence of normal environmental clues. Recent work suggests that the differentiation of rods may be more complex and that the differentiation of these cells might be modified by environmental clues (see Raymond, this volume).

The simple possibilities for "determination of address" in the retina are the following. First, cells could be committed to both a highly specific final differentiation fate (amacrine cell, using substance P as a neurotransmitter and enkephalin as a neuromodulator) and committed to recognizing a particular address determinant. These cells would migrate, stop migrating and differentiate as independent decisions. The second simple possibility is that cells could be committed only to recognizing a particular address marker. They would then differentiate after reaching that address according to signals from the local environment. These cells would make only two independent decisions, migrate and stop migrating at a particular place and time.

It should be clear that we do not yet know which of these possibilities is the case for any given retinal neuron. Likely the answer will be highly dependent on what level of cell differentiation is being considered. The decision to become a ganglion cell might be made at a different time and in a different way from the decision as what neurotransmitter to use. The recent work on the developmental pathways taken by photoreceptors promises to shed light on these questions.

Unfortunately, we know very little about other cell types in the retina besides the photoreceptors. Partially, this is a problem of cell-type identification. Often the most telling characteristic of the cell is its shape. However, it is not clear if that shape is a consequence of where the cell finds itself or of some intrinsic tendency to form that shape. It would seem at this juncture that size and shape might be found to be quite different in their developmental control.

Much recent work has focused on what seems to be a relatively late differentiation decision for retinal neurons, the acquisition of specific transmitter characteristics. These studies are still at the descriptive stage and little developmental knowledge is available. However, the result is already a bewildering array of potential differentiated cell types, clearly more than 50-70 in the retina (Marc et al., 1978; Eldred et al., 1987). However, there are recognizable patterns and certain transmitters or modulators such as enkephalin seem to be associated with a single class of neuron such as amacrine cells. Thus, the eventual appearance of enkephalin could be the start of an investigation of the more covert (and presumably earlier) decision to become a retina amacrine cell.

POSSIBLE ROLE OF COGNIN IN THIS CASCADE OF DECISIONS

In our current understanding, cognin in the retina seems associated with cholinergic cell types. What cells are these? Those most widely accepted are the cells which communicate between the retina and the rest of the CNS, the retina ganglion cells (Puro, 1985). Cognin is clearly on the surface of these cells throughout its appearance (Dobi et al., 1986, 1988) (Figure 3).

However, there are other cholinergic cells in the retina which (at certain times of development and in certain species) might be both pre and post-synaptic to ganglion cells (Sakai et al., 1986). These are the other cell type which shares the GCL as a residence, the displaced amacrine cells. These are not rare cells, but might constitute up to 70% of the retina in some species (Wassle et al., 1987). These cells also exhibit cognin during these stages of differentiation (figure 3).

To investigate these cholinergic cells, we sought an additional marker. We used choline acetyltransferase (ChAT) both because displaced amacrines are often the presynaptic cells where this enzyme might be expected and because we already had information about its developmental kinetics. This enzyme increases in the

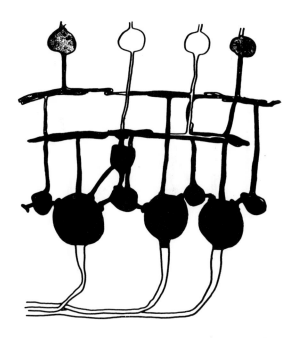

Figure 4. Schematic drawing of the cells bordering the inner plexiform layer (IPL) of the late embryonic chick neural retina. The distribution of choline acetyltransferase (ChAT) or ChAT-like immunoreactivity between embryonic day 15 and embryonic day 19 is shown. Cells are described as in the legend to Figure 1. Briefly, some amacrine, most displaced amacrine and most ganglion cell bodies are stained.

cells both in the tissue and in culture as cognin on the cell surface declines (Hausman et al., 1986). Recently, Miles Epstein and is collaborators have developed antibodies to chicken ChAT (Johnson and Epstein, 1986; Millar et al., 1985; 1987). We have used these in colocalization studies with cognin and found an intriguing developmental correlation. ChAT (or ChAT-like activity) seems to be present in the early ganglion and displaced amacrine cells (Figure 4), but is subsequently lost from the ganglion cells and the adult (displaced amacrine cell only) pattern is seen (Dobi et al., 1988) (Figure 5). A similar transient appearance of ChAT-like immunologic activity might be a general property of CNS development (Gould and Butcher, 1987).

The development of cholinergic neurons is complex and is known to be affected by various substances such as NGF (Martinez et al., 1987) and tissue extracts (Iacovitti et al., 1987). Puro had noted

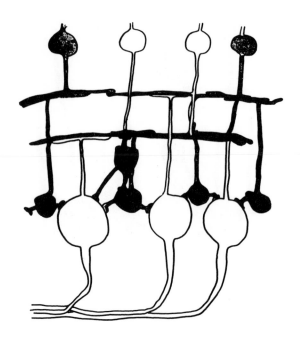

Figure 5. Schematic drawing of the cells bordering the inner plexiform layer (IPL) of the late embryonic chick neural retina. The distribution of choline acetyltransferase (ChAT) or ChAT-like immunoreactivity after embryonic day 19 is shown. Cells are as described in the legend to Figure 1. By comparison to Figure 4, the ganglion cells have lost staining.

several years ago that insulin potentiated the precocious induction of acetylcholine release in cultures of retina neurons (Puro and Agardh, 1984). Thus, another marker for cholinergic differentiation might be the ability to specifically bind insulin. Because of the association of cognin with cells undergoing cholinergic development, we were interested in Puro's observation and have followed up on it. While specific cell types have not yet been identified, insulin clearly binds to chick embryo retinal neurons (Peterson et al., 1986). This binding is developmentally-dependent and reaches a maximum at the same time as the change in cognin distribution. There is a dose-dependent correlation between the amount of insulin present during this period of peak binding and subsequent cholinergic differentiation of retina neurons (measured by increases in ChAT activity)(Kyriakis et al., 1987). As yet, little is known about the possible role of cognin in this insulin-mediated differentiation.

The current state of our understanding of cognin is the hypothesis that it might act as an address marker for cells associated with cholinergic differentiation in the embryonic chick retina. The challenge is how to test that hypothesis. We need to know more about the ontogeny of the cells which take up residence in the chick retina GCL (Figure 3). It would be useful to know when these cells first start synthesizing cognin and how this synthesis correlates in development with cessation of mitosis and cell migration. Cognin protein disappears from the cell surfaces only after the specification of neurotransmitters appears to be decided (Meller, 1984). However, we do not know when cognin synthesis ceases and with which events this might be correlated. To answer these questions we will need probes for cognin mRNA and are currently moving in that direction.

Since cognin binds to a qualitatively different protein, the 64 kd cognin-binding protein, knowledge of the control of cognin expression can only provide half the story. It will be critical to obtain information about the tissue and cellular distribution of the 64 kd protein and eventually information on its synthesis and turnover.

REFERENCES

Adler, R., Lindsey, J.D. and C.L. Elsner. 1984. Expression of cone-like properties of chick embryo neural retina cells in glial-free monolayer culture. J. Cell Biol. 99:173-1178.

Baughman, R.W. and C.R. Bader. 1977. Biochemical characterization and cellular localization of the cholinergic system in the chicken retina. Brain Res. 138:469-485.

Ben-Shaul, Y., Hausman, R.E. and A.A. Moscona. 1979. Visualization of a cell surface glycoprotein, the retina cognin, on embryonic cells by immuno-latex labeling for scanning electron microscopy. Develop. Biol. 72:89-101.

Dobi, E.T., Troccoli, N.M. and R.E. Hausman. 1986. Distribution of R-cognin in late embryonic and post-hatching chick retina. Invest. Ophthal. and Visual Sci. 27:323-329

Dobi, E.T., Naya, F.J. and R.E. Hausman. 1988. Distribution of R-cognin and choline acetyltransferase in the ganglion cell layer of developing chick neural retina. Cell Different. 22:115-124.

Eldred, W.D., Li, H-B., Carraway, R.E. and J.E. Dowling. 1987. Immunocytochemical localization of LANT-6-like immunoreactivity within neurons in the inner nuclear and ganglion cell layers in vertebrate retinas. Brain Res. 424:361-370.

Fujita, S. and M. Horii. 1963. Analysis of cytogenesis in chick retina by tritiated thymidine autoradiography. Arch. Hist. Jap. 23:359-366.

Garber, B.B. and A.A. Moscona. 1972. Reconstruction of brain tissue from cell suspensions. II. Specific enhancement of aggregation of embryonic cerebral cells by supernatant from homologous cell cultures. Develop. Biol. 27:235-271.

Gould, E. and L.L. Butcher. 1987. Transient expression of choline acetyltransferase-like immunoreactivity in Purkinje cells of the developing rat cerebellum. Develop. Brain Res. 34:303-306.

Hausman, R.E. and A.A. Moscona. 1973. Cell surface interactions: Inhibition by proflavine of embryonic cell aggregation and the production of specific cell aggregating factor. Proc. Nat. Acad. Sci. USA 70:3111-3114.

Hausman, R.E. and A.A. Moscona. 1975. Purification and Characterization of the neural retina cell aggregating factor. Proc. Nat. Acad. Aci. USA 72:916-920.

Hausman, R.E. and A.A. Moscona. 1976. Isolation of retina-specific cell aggregating factor from membranes of embryonic retina tissue. Proc. Nat. Acad. Sci. USA 73:3594-3598.

Hausman, R.E. and A.A. Moscona. In vitro studies on embryonic cell associations. In "Tests of Teratogenicity in vitro" eds. Ebert, J.D. and Marois, M., North Holland, Amsterdam (1976), pp. 171-185.

Hausman, R.E., Knapp, L.W. and A.A. Moscona. 1976. Preparation of tissue-specific cell aggregating factors from embryonic neural tissues. J. Exptl. Zool. 198:417-422.

Hausman, R.E. and A. A. Moscona. 1979. Immunologic detection of retina cognin on the surface of the embryonic cells. Exptl. Cell Res. 119:191-204.

Hausman, R.E., Christie, T., Gliniak, B.C. and W.A. Rosenkrans. 1985. Topological correlation between the cell-recognition protein, R-cognin, and α-bungarotoxin receptor in retinal plasma membrane. Int. J. Develop. Neurosci. 3:41-50.

Hausman, R.E., Katz, M.S., Dobi, E.T. and J. Offermann. 1986. Cognin distribution during differentiation of embryonic chick retinal cells in vitro. Int. J. Develop. Neurosci. 4:537-544.

Hausman, R.E., Dobi, E.T., Troccoli, N.M. and T. Christie. 1987. Possible roles for cognin in retina development. American Zool. 27:171-178.

Hoffman, S., Sorkin, B.C., White, P.C., Brackenbury, R., Mailhammer, R., Rutishauser, U., Cunningham, B.A. and G.M. Edelman. 1982. Chemical characterization of a neural cell adhesion molecule purified from embryonic brain membranes. J. Biol. Chem. 257:7720-7729.

Iacovitti, L., Teitelman, G., Joh, T.H. and D.J. Reis. 1987. Chick eye extract promotes expression of a cholinergic enzyme in sympathetic ganglia in culture. Develop. Brain Res. 33:59-65.

Jacobson, M. 1984. Cell lineage analysis of neural induction: Origins of cells forming the induced nervous system. Develop. Biol. 102:122-129.

Johnson, C.D. and M.L. Epstein. 1986. Monoclonal antibodies and polyvalent antiserum to chicken choline acetyltransferase. J. Neurochem. 46:968-976.

Kahn, A.J. 1974. An autoradiographic analysis of the time of appearance of neurons in the developing chick neural retina. Develop. Biol. 38:30-40.

Kyriakis, J.M., Hausman, R.E., and S.W. Peterson. 1987. Insulin stimulates choline acetyltransferase activity in cultured embryonic chick retinal neurons. Proc. Nat. Acad. Sci. USA 84:7463-7467.

Lilien, J.E. 1968. Specific enhancement of cell aggregation in vitro. Develop Biol. 17:658-678.

Lilien, J.E. and A.A. Moscona. 1967. Cell aggregation: its enhancement by a supernatant from cultures of homologous cells. Science 157:70-72.

Marc, R.E., Stell, W.K., Bok, D. and D.M.K. Lam. 1978. GABAergic pathways in the goldfish retina. J. Comp. Neurol. 182:221-246.

Marchase, R.B. 1987. Glucose-1-phosphate containing glycoproteins and interneuronal adhesion. Amer. Zool. 27:179-187.

Martinez, H.J., Dreyfus, C.F., Miller Jonakait, G. and I.B. Black. 1987. Nerve growth factor selectively increases cholinergic markers but not neuropeptides in rat basal forebrain in culture. Brain Res. 412:295-301.

McClay, D.R. and A.A. Moscona. 1974. Purification of the specific cell-aggregating factor from embryonic neural retina cells. Exptl. Cell Res. 87:438-443.

Meller, K. 1984. Morphological studies on the development of the retina. In: Progress in Retinal Research, Vol 3: eds. N.N.Osborne and G.J. Chader (Pergamon Press, New York) pp. 1-19.

Millar, T., Ishimoto, L., Johnson, C.D., Epstein, M.L., Chubb, I.W. and I.G.Morgan. 1985. Cholinergic and acetylcholinesterase-containin neurons of the chicken retina. Neurosci. Lets. 61:311-316.

Millar, T., Ishimoto, L., Boelen, M., Epstein, M.L., Johnson, C.D. and I.G.Morgan. 1987. The toxic effects of ethylcholine mustard aziridinium ion on cholinergic cells in the chicken retina. J. Neurosci. 7:343-356.

Moscona, A. 1956. Development of heterotypic combinations of dissociated embryonic chick cells. Proc. Soc. Exptl. Biol. Med. 92:410-416.

Moscona, A. 1957. Formation of lentoids by dissociated retinal cells of the chick embryo. Science 125:598-599.

Moscona, A. 1961. Rotation-mediated histogenetic aggregation of dissociated cells: a quantifiable approach to cell interactions in vitro. Exptl. Cell Res. 22:455-475.

Moscona, A.A. 1962. Analysis of cell recombinations in experimental synthesis of tissues in vitro. J. Cell. Comp. Physiol. 60:suppl. 65-80.

Moscona, A.A. 1962. Cellular interactions in experimental histogenesis. Intern. Rev. Exptl. Path. 1:371-428.

Moscona, A.A. 1963. Studies on cell aggregation: demonstration of materials with selective cell-binding activity. Proc. Nat. Acad. Sci. USA 49:742-747.

Moscona, A.A. 1968. Properties of specific cell ligands and their role in the formation of multicellular systems. Develop. Biol. 18:250-277.

Moscona, A.A. and R.E. Hausman. Biological and biochemical studies on embryonic cell recognition. In: "Cell and Tissue Interactions", eds. Lash, J.W. and Burger, M.M., Raven Press, New York (1977). pp. 173-185.

Ophir, L., Moscona, A.A. and Y. Ben-Shaul. 1984. Cell disorganization and malformation in neural retina caused by antibodies to R-cognin: ultrastructural study. Cell Different. 15:53-59.

Peterson, S.W., Kyriakis, J.M. and R.E. Hausman. 1986. Changes in insulin binding to developing embryonic chick neural retina cells. J. Neurochem. 47:851-855.

150

Puro, D.G. 1985. Cholinergic systems. In: Retinal transmitters and modulators: Models for the brain. ed. W.W. Morgan (CRC Press, Boca Raton, FL) pp.63-92.

Puro, D.G. and E. Agardh. 1984. Insulin-mediated regulation of neuronal maturation. Science 255:1170-1172.

Rutishauser, U. 1984. Developmental biology of a neural cell adhesion molecule. Nature 310:549-554.

Rutishauser, U., Hoffman, S. and G. Edelman. 1982. Binding properties of a cell adhesion molecule from neural tissue. Proc. Natl. Acad. Sci. USA 79:685-689.

Sakai, H.M., Naka, K-I. and J.E. Dowling. 1986. Ganglion cell dendrites are presynaptic in catfish retina. Nature 319:495-497.

Schmitt. M., Painter, R.G., Jesitis, A.J., Preissner, K., Sklar, L.A., Schubert, D. and M. LaCorbiere. 1985. Isolation of an adhesion-mediating protein form chick neural retina adherons. J. Cell Biol. 101:1071-1077.

Sidman, R.L. 1975. Cell interaction in mammalian brain development. In: The Nervous System, Vol 1: The Basic Neurosciences, ed. D.B. Tower (Raven Press, New York) pp.601-610.

Thiery, J-P., Brackenbury, R., Rutishauser, U. and G.M. Edelman. 1977. Adhesion among neural cells of the chick embryo. II. Purification and characterization of a cell adhesion molecule from neural retina. J. Biol. Chem. 252:6841-6845.

Troccoli, N.M. and R.E. Hausman. 1985. Vesicle interactions as a model for the retinal cell-cell recognition mediated by R-cognin. Cell Different. 16:43-49.

Troccoli, N.M. and R.E. Hausman. 1988. R-cognin does not bind to itself during membrane interaction in vitro. Cell Different. 22:225-232.

Vogel, Z. and M. Nirenberg. 1976. Localization of acetylcholine receptors during synaptogenesis in retina. Proc. Nat. Acad. Sci. USA 73:1806-1810.

Vogel, Z., Maloney, G.J., Ling, A. and M.P. Daniels. 1977. Identification of synaptic acetylcholine receptor sides in the retina with peroxidase-labeled alpha bungarotoxin. Proc. Nat. Acad. Sci. USA 74:3268-3272.

Wassle, H., Chuh, M.H. and F. Muller. 1987. Amacrine cells in the ganglion cell layer of the cat retina. Invest. Ophthal. Vis. Sci. 28:suppl. 261.

Embryonic Chick Neural Retinal Cell Interactions with Extracellular Matrix Proteins: Characterization of Neuronal ECM Receptors and Changes in Receptor Activity During Development

Deborah E. Hall, Karla M. Neugebauer, and Louis F. Reichardt

THE ROLE OF EXTRACELLULAR MATRIX (ECM) IN RETINAL GANGLION CELL AXON OUTGROWTH.

Retinal ganglion cells are born and initiate axon growth in the most vitreal layer of the developing neural retina (Rager, 1980; Krayanek and Goldberg; 1981). In this environment axonal growth cones encounter three different substrates: the inner limiting membrane (a basal lamina), Müller glial cells, and the axons of other neurons (Easter et al., 1984). Both astrocytes and axons have been shown to be excellent substrates for many neurons in vitro, including retinal ganglion cells (McCaffery et al., 1984; Cohen et al., 1986; Fallon, 1985; Tomaselli et al., 1986; Chang et al., 1987; Kapfhammer and Raper, 1987). Protein constituents of the extracellular matrix (ECM) such as laminin can also serve as substrates for neurite growth by neural retinal cells (Manthorpe et al., 1983; Rogers et al., 1983; Smallheiser et al., 1984; Adler et al., 1985). The inner limiting membrane has been shown to contain laminin and other ECM proteins throughout development (Jerdan et al., 1984; McLoon et al., 1984; Adler et al., 1985; Halfter and Chen, 1987). During the first week of development, laminin immunoreactivity is not confined to the inner limiting membrane but is present around glial endfeet and within the optic stalk and optic nerve (Cohen et al., 1987). Since ganglion cell axons are initiated during this period, laminin is temporally and anatomically well placed to influence retinal ganglion cell axon outgrowth.

Experiments using cultured explants of embryonic retina provide more direct evidence for the role of the inner limiting membrane in promoting and orienting axonal outgrowth. First, the isolated inner limiting membrane has been shown to be an excellent substrate for neurite growth in vitro (Halfter et al., 1987). Second, process outgrowth by retinal ganglion cells in explanted retinal whole mounts is inhibited by antibodies that prevent neuronal interactions with the ECM (Henke-Fahle and Bonhoeffer, 1983; Halfter and Deiss, 1986). Enzymatic removal of the inner limiting membrane disrupts orderly axon growth to the optic disc, although it does not prevent process outgrowth in explanted retinal whole mounts (Halfter and Deiss, 1984). Thus, it seems that integrity of the retinal explants, including cues provided by the inner limiting membrane, is required for the guidance of retinal ganglion axon growth. The isolated inner limiting membrane, though, does not contain directional cues for process growth in vitro (Halfter et al., 1987), so it is not yet clear whether cells, axons, or the ECM play the major role in orienting these fibers. To summarize, the ECM appears to have a prominent function in promoting the extension of retinal ganglion axons. Individual components of the ECM that promote neurite outgrowth in vitro, most notably laminin, seem likely to have a similar function in vivo wherever they are found along the routes taken by retinal axons.

THE SUBSTRATE REQUIREMENTS FOR NEURAL RETINAL CELL ATTACHMENT AND NEURITE OUTGROWTH.

As illustrated in Fig. 1, neuron/substrate interactions involve a complex series of events including cell attachment, cell spreading, lamellipodia formation, neurite initiation and finally neurite extension (Bray and Chapman, 1985; Goldberg and Burmeister, 1986; DeGeorge et al., 1985). These events may be mediated by individual proteins or combinations of proteins in the ECM. Basement membranes share several prominent components, including laminin, entactin, collagen IV and heparan sulfate proteoglycans. Since they do vary in structure and composition (cf. Mohon and Spiro, 1986; Chiu et al., 1986), different ECMs may well differ in their capacity to support neurite outgrowth.

To examine the roles of extracellular matrix components and neuronal cell surface receptors, we examined cell attachment and neurite outgrowth in separate quantitative bioassays. These are diagrammed in Figure 2 and have been described in detail elsewhere (Hall et al., 1987). Briefly, the cell attachment assay measured

NEURON/EXTRACELLULAR MATRIX INTERACTION

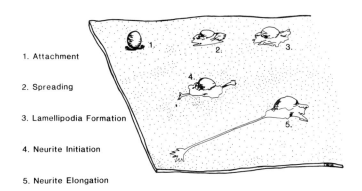

1. Attachment

2. Spreading

3. Lamellipodia Formation

4. Neurite Initiation

5. Neurite Elongation

Figure 1. Neuron/Extracellular Matrix Interaction. Different phases of neuron/matrix interaction, as described by DeGeorge et al. (1985), are diagrammed. (1.) Cell attachment involves adherence of neurons to a substrate but does not involve any cytoskeletal changes. (2.) Spreading involves both more extensive interaction with the matrix and cytoskeletal changes that result in a major cellular shape change. (3.) Lamellipodia formation involves extension of a ruffling membrane structure, the lamellipodium, which is rich in actin filaments, does not contain microtubules and is a prerequisite to neurite initiation. Treatment with colchicine, which disrupts microtubules, has no effect upon attachment or lamellipodium formation (DeGeorge et al., 1985). (4. and 5.) Neurite initiation and elongation require both protein synthesis and intact microtubules (DeGeorge et al., 1985).

the number of neural cells that attached to different substrates during a 1.5 h incubation period relative to positive (poly-D-lysine-coated) and negative (bovine serum albumin-coated) controls. To measure neurite outgrowth, neural cells were grown overnight and were then scored for the proportion of cells bearing neurites longer than two cell diameters. Antibodies to different matrix proteins and antibodies to cell surface glycoproteins were used to determine the role of individual molecules in mediating neuronal interactions with different substrates.

Neural retinal cells attach well to laminin, collagen IV and fibronectin substrates (Fig. 3, top panel). Attachment to all three substrates was concentration-dependent. Fibronectin was somewhat

154

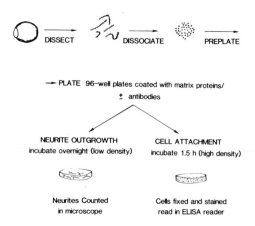

DISSECT → DISSOCIATE → PREPLATE

→ PLATE 96-well plates coated with matrix proteins/
± antibodies

NEURITE OUTGROWTH
incubate overnight (low density)

CELL ATTACHMENT
incubate 1.5 h (high density)

Neurites Counted
in microscope

Cells fixed and stained
read in ELISA reader

Figure 2. Diagram of Cell Attachment and Neurite Outgrowth Bioassays. Neural retinal cells are obtained from embryonic chick retinae and enzymatically dissociated with trypsin. After they are preplated on tissue culture plastic to remove non-neuronal cells, neural cells are plated onto 96 well plates that have been previously coated with purified extracellular matrix proteins or other test substrates and blocked with 1 % bovine serum albumin. Plates to be used for cell attachment assays are incubated at 37°C for 1.5 h. Then unattached cells are removed and attached cells are fixed overnight. Attachment is quantitated by staining the cells with trypan blue and then measuring A_{690} in an ELISA plate reader. Cell attachment data are expressed as percent of cells attached relative to positive (1 mg/ml poly-D-lysine-coated) and negative (bovine serum albumin-coated) control wells. Plates to be used for neurite outgrowth assays are incubated overnight at 37°C, fixed and scored for percent of cells bearing neurites by examination in the microscope. For experiments examining the effects of antibodies, antibodies were added to the wells prior to cell plating at concentrations as noted.

less effective as a substrate for cell attachment than either collagen IV or laminin: even at high coating concentrations of fibronectin (300 µg/ml) only about 70 % of the cells attached. This relatively low affinity of retinal neurons for fibronectin corroborates previous work (Rogers et al., 1983). Cell attachment was substrate specific since anti-laminin antibodies blocked attachment to laminin but not to collagen IV, and anti-collagen IV antibodies blocked attachment to collagen IV but not to laminin (Hall et al, 1987).

Among ECM proteins, laminin is distinctive in its ability to promote rapid neurite outgrowth. A very high proportion of neural retinal cells grew neurites on laminin and neurite outgrowth on

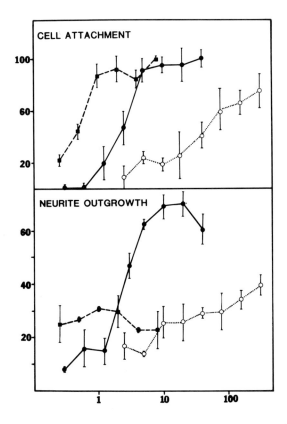

Figure 3. Quantitation of Cell Attachment and Neurite Outgrowth by Embryonic (E6) Neural Retinal Cells on Extracellular Matrix Protein Substrates. Cell attachment (top panel) and neurite outgrowth (bottom panel) on laminin (solid line), fibronectin (dotted line) and collagen IV (dashed line) are shown. E6 chick retinal cells were plated on tissue culture wells coated with either laminin, fibronectin, or collagen IV and maintained in serum-free medium for either 1.5 h (attachment) or 16-24 h (neurite outgrowth). Each point is the average of at least three separate determinations. The figure shows the percent of cells attached (top panel) or the percent of cells with neurites (bottom panel) vs. protein coating concentration in μg/ml. The error bars show standard error of the mean.

laminin was concentration-dependent (Fig. 3, bottom panel). In contrast, a low and fairly constant level of neurite outgrowth was seen on a range of concentrations of both collagen IV and fibronectin (Fig. 3). In addition to the quantitative differences between levels of neurite outgrowth seen on laminin and the other substrates, there were qualitative differences as well. Neurons appeared well attached on all three substrates and had a characteristic polygonal shape. Neurites on laminin were long, branched and usually unfasciculated. In contrast, neurites growing on collagen IV and fibronectin were frequently fasciculated, shorter and usually unbranched (Hall et al., 1987).

The above differences in the substrate requirements for cell attachment and neurite outgrowth suggest that adherence to a substrate is a prerequisite, but is not sufficient, for neurite outgrowth. First, our data show that laminin is unique among ECM proteins in promoting profuse neurite outgrowth by neuroretinal cells, despite attachment with equal or greater affinity to collagen IV and poly-D-lysine. In other examples of this dichotomy, Gundersen (1987) has shown that chick sensory neurons extend neurites on laminin in preference to collagen IV, even though neurites are more strongly bound to the latter. Perhaps the most dramatic example of laminin's strong neurite-promoting activity is its ability to promote neurite outgrowth in the absence of NGF by sympathetic and sensory neurons, which extend virtually no neurites on other ECM substrates without NGF, even though they attach well to these substrates (cf. Lander et al., 1983). Finally, it has been possible to inhibit neurite outgrowth on several cellular substrates with antibodies to ECM receptors and cell adhesion molecules without reducing neuronal attachment (cf. Bixby et al., 1987; Tomaselli et al., 1988; Bixby et al., 1988; Neugebauer et al., 1988). These results do not support a strict correlation between adhesive strength and a substrate's ability to promote neurite outgrowth.

LAMININ STRUCTURE.

Do distinct neuronal responses to laminin reflect something unique about the structure of laminin and/or the neuronal cell surface molecules that mediate interactions with laminin? To begin to address this question we have studied how neurons interact with different domains of the laminin molecule and have attempted to determine what neuronal surface molecules mediate this interaction. Laminin is a large glycoprotein (Mr 10^6) composed of three different subunits (Timpl et al, 1979; Fig. 4). Laminin binds to collagen, heparan sulfate proteoglycans, and cells; these functions have been ascribed to separate structural domains of laminin (Timpl et al., 1983; Charonis et al., 1985; Rao et al., 1982; Terranova et

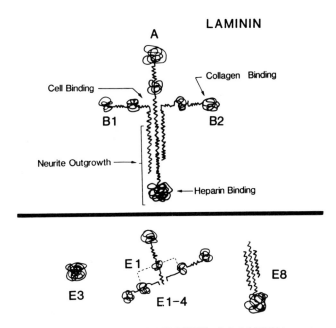

LAMININ

ELASTASE FRAGMENTS of LAMININ

Figure 4. Structural Diagram of Laminin and Elastase Fragments of Laminin. The three dissimilar subunits (A, B1, and B2) and the collagen, cell and heparin-binding sites of laminin are shown. Both the B1 and B2 chains are thought to have collagen binding domains; only one is indicated. Experimental evidence suggests that cells bind to two distinct domains, one near the intersection of the cross and the other near the end of the long arm (Rao et al., 1982; Terranova et al., 1983; Aumailley et al., 1987; Goodman et al., 1987; Graf et al., 1987). A domain which promotes neurite outgrowth is near but apparently not included within fragment E3 (Edgar et al., 1984; Engvall et al., 1987).

al., 1983; Aumailley et al., 1987; Goodman et al., 1987; Graf et al., 1987; Fig. 4). Three types of evidence suggest that a domain near the heparin-binding domain (E3) mediates neuronal interaction with laminin. First, anti-E3 antibodies block neurite outgrowth on whole laminin though E3 substrates do not promote neurite outgrowth (Edgar et al, 1984). Secondly, fragment E8 which contains the heparin-binding domain (E3) and a portion of the alpha helical region of the long arm of laminin promotes neurite outgrowth by chick sympathetic neurons (Edgar et al, 1984). Neurite outgrowth in response to laminin domain E8 is likewise inhibited by anti-E3 antibodies (Edgar et al, 1984). Lastly, monoclonal antibodies that react with the laminin molecule near the E3 domain inhibit neurite outgrowth on whole laminin (Engvall et al, 1987). In addition to

laminin fragments E3 and E8, fragments E1 and E1-4 (see Fig. 4) are
of particular interest since they have been shown to mediate
attachment and spreading of a number of different cell types (Timpl
et al.,1983; Rao et al., 1982; Terranova et al., 1983; Graf et al.,
1987; Goodman et al., 1987; Aumailley et al., 1987; Fig. 4).

To begin to understand the details of how retinal neurons
interact with laminin, biologically active proteolytic fragments of
the laminin molecule were generated by elastase digestion as
previously described (Ott et al, 1982; for a detailed description of
the laminin fragment preparation see Hall et al, in preparation).
Anti-E3 antibodies were prepared by immunization of rabbits with
purified laminin fragment E3 by standard protocols.

Laminin fragment E8, but not fragment E1-4, promoted retinal
neuron attachment (Hall et al, in preparation) and neurite outgrowth
(Fig. 5). Even after overnight in culture, neural retinal cells
were not attached to the E1-4 substrate and formed cell aggregates
that floated above the culture substrate. Both attachment and
neurite outgrowth on fragment E8 were concentration-dependent (Hall
et al, in preparation). E3 did not support attachment and neurite
outgrowth (data not shown). However, anti-E3 antibodies blocked both
cell attachment and neurite outgrowth by neural retinal cells on
whole laminin as well as on fragment E8 (Fig. 5 and Hall et al.,
1988). Thus, as previously shown for sympathetic neurons (Edgar et
al., 1984), laminin fragment E8 appears to be sufficient to promote
neurite outgrowth for neural retinal cells.

NEURONAL CELL SURFACE MOLECULES OF THE INTEGRIN β_1 RECEPTOR FAMILY
MEDIATE INTERACTIONS WITH EXTRACELLULAR MATRIX PROTEIN SUBSTRATES
AND WITH LAMININ FRAGMENT E8.

Cell surface receptors that mediate cell interactions with a
variety of different extracellular matrix proteins have been
identified as members of a superfamily of proteins, the integrins
(reviewed in Hynes, 1987 and Ruoslahti and Pierschbacher, 1987).
These proteins are heterodimeric membrane-spanning glycoproteins
with distinct α and β subunits. Three distinct β subunits (β_1
through β_3) have been identified to date. Integrins containing two
of these (β_1 and β_3) have been shown to mediate cellular
interactions with ECM proteins. β_3 class receptors appear to mediate
attachment to vitronectin, fibrinogen, von Willebrand factor, and

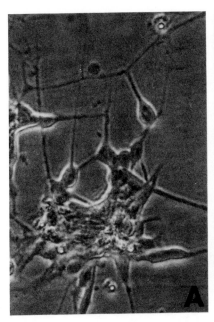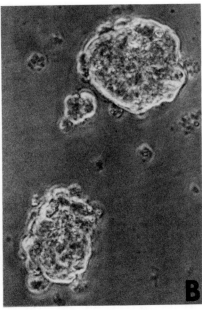

Figure 5. Photographs of Embryonic Neural Retinal Cells Grown Overnight on Laminin Fragment E8 in the Presence and Absence of Anti-E3 Antibody. Embryonic neural retinal cells (E6) were grown overnight on laminin fragment E8 in the absence (A) and presence (B) of anti-E3 antiserum (1/100 dilution). Anti-E3 antiserum blocks cell attachment and neurite outgrowth on whole laminin and on laminin fragment E8 (see also Edgar et al., 1984).

fibronectin (Pytela et al., 1985b; 1986; Hynes, 1987). β_1 class receptors appear to mediate interactions with fibronectin, laminin, and collagens I and IV (Horwitz et al., 1985; Hall et al., 1987). Within a receptor family sharing a common β subunit, the α subunit appears to confer substrate specificity (Ruoslahti and Pierschbacher, 1987). Several of these α subunits have been identified and implicated in binding to specific substrates, including fibronectin, vitronectin, fibrinogen and von Willebrand Factor (Pytela et al., 1985a; 1985b, 1986). In other cases, such as laminin and collagen IV, the specific α subunits have not been identified, but the role of integrin - β_1 class receptors has been inferred from the inhibitory effects on attachment of anti-integrin β_1 antibodies. Thus, β_1 subunit-specific antibodies such as CSAT (Buck et al., 1986), prevent attachment of many cells to laminin, fibronectin, and collagens I and IV (Horwitz et al., 1985; Bozyzcko and Horwitz, 1986; Hall et al., 1987). Hence, the receptors for each

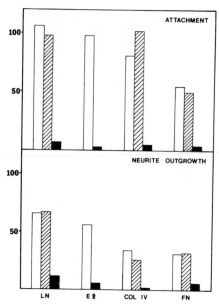

Figure 6.   Effect of Anti-Integrin β_1 Antibodies (CSAT) Upon Cell
Attachment and Neurite Outgrowth.   Cell attachment (top panel) and
neurite outgrowth (bottom panel) by E6 neural retinal cells was
measured as above and is expressed as percent of cells attached
(top) or percent of cells with neurites (bottom).   Cell interaction
with different matrix protein substrates in the presence of CSAT (5-
10 µg/ml; black bars), a control murine IgG (cross hatched bars) or
no antibody (open bars) is shown.   Wells were coated with 5 µg/ml
laminin, 16 µg/ml E8, 2 µg/ml collagen IV or 300 µg/ml fibronectin.

of these ECM glycoproteins are thought to be heterodimers containing
the β_1 subunit.

The anti-integrin β_1 antibody, CSAT, blocks both attachment and
neurite outgrowth by neural retinal cells on fibronectin, laminin
and collagen IV, indicating that integrin receptors mediate neuronal
interactions with these three substrates (Fig. 6).   Neural retinal
cells grown overnight on a laminin substrate in the presence of the
anti-integrin β_1 antibody were not attached to the substrate, but
floated in large aggregates instead (Fig. 7).   In addition, the
antibody inhibited neural retinal cell attachment and neurite
outgrowth on laminin fragment E8 suggesting that E8 contains a
binding site recognized by an integrin receptor (Fig. 6).

Since anti-integrin β_1 eliminates attachment to these ECM
protein substrates, integrin receptors may mediate cell attachment,
a prerequisite for neurite outgrowth, but may not be directly
involved in neurite outgrowth.   Two experiments address this

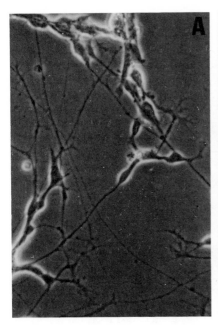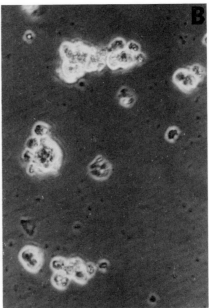

Figure 7.   Photographs Showing the Effects of the Anti-Integrin β_1 Antibody (CSAT) Upon Neurite Outgrowth on Laminin.   Embryonic chick neural retinal cells (E6) were cultured overnight on a laminin substrate (coated with 5 µg/ml) in the absence (A) or presence (B) of CSAT (10 µg/ml).   In the presence of the CSAT antibody neurons do not attach to the substrate even after overnight in culture and aggregate into small balls of cells.   Incubation with metabolic dyes indicated that these cell aggregates were alive and that antibody effects were not due to cytotoxicity (Hall et al., 1987).

question directly.   In the first, neural retinal cells were grown on a substrate consisting of the anti-integrin β_1 antibody itself.   As shown in Fig. 8, this antibody substrate promoted cell attachment and neurite outgrowth in a concentration-dependent fashion, much the same as laminin does.   In contrast, polyclonal antibodies against N-CAM, a neuronal surface molecule, (Hoffman et al., 1982), promoted cell attachment but not neurite outgrowth (Hall et al., 1987).   In the second approach, ciliary neurons were grown on mixed substrates of poly-D-lysine and laminin (Tomaselli et al., 1986).   Under these conditions, integrin β_1 antibody did not inhibit attachment to the polycationic substrate, yet neurite outgrowth was eliminated.   Thus, under conditions where neurons are well attached to the substrate, anti-integrin β_1 still inhibits laminin-dependent neurite outgrowth (Tomaselli et al., 1986).   These experiments indicate that integrin

162

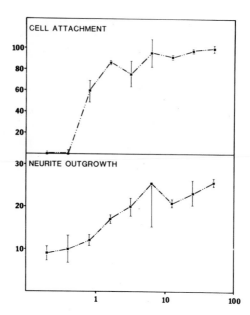

Figure 8. Quantitation of Neural Retinal Cell Attachment and Neurite Outgrowth on a CSAT Antibody Substrate. Cell attachment (top panel) and neurite outgrowth (bottom panel) on CSAT antibody coated wells were measured as for matrix protein substrates. E6 neural retinal cells were plated onto wells coated with different amounts of the CSAT antibody. Percent of cells attached (top panel) and percent of cells with neurites (bottom panel) vs. protein coating concentration in µg/ml are shown. Note that both cell attachment and neurite outgrowth are concentration dependent on the antibody substrate. The error bars show standard error of the mean.

receptors play a direct role in both neuron attachment to ECM substrates and in ECM-dependent neurite outgrowth. The presence of integrin β_1 antigen on neurites and growth cones as well as on cell bodies (Fig. 9; Bozyczko and Horwitz, 1986) and the ability of integrin β_1 antibodies to detach existing neurites from laminin substrates (Bozyczko and Horwitz, 1986) are also consistent with a direct role of integrins in neurite extension.

NEURAL RETINAL CELL SUBSTRATE PREFERENCE CHANGES DURING DEVELOPMENT.

The environment encountered by chick retinal ganglion cell axons as they grow toward their target, the optic tectum, changes during development. Within the eye, growth cones appear to contact three distinct substrates; the inner limiting membrane (a basal lamina), Müller glial endfeet and retinal ganglion cell axons. As

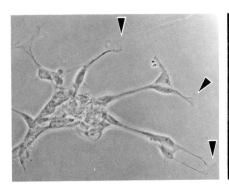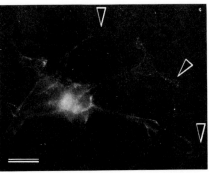

Figure 9. Localization of the Integrin β_1 Subunit on the Surface of E6 Retinal Neurons. Phase contrast (left panel) and immunofluorescence (right panel) microscopic images of retinal neurons are shown. Both cell bodies and neurites are labeled by the antibody, but the staining is noticeably weaker on the growth cones (arrows). E6 retinal neurons were plated on coverslips coated with 10 µg/ml laminin and cultured overnight. After fixation with 100% methanol at -20°C for 3 min., the cells were stained with an integrin β_1 polyclonal antibody (25 mg/ml) followed by a rhodamine-conjugated second antibody. Scale bar: 20 mm.

development proceeds, retinal ganglion cell axons exit the eye through the optic fissure, and their growth in the optic tract proceeds along neuroepithelial glial endfeet as well as the surfaces of preexisting axons (Rager, 1980; Silver and Sapiro, 1981; Silver, 1984). As the substrates contacted by growth cones growing within the optic nerve differ from those within the eye, it seems possible that neuronal substrate preferences would change with development. To examine this possibility, neuroretinal cells were isolated from embryos of different ages and tested for their ability to interact with purified ECM proteins in cell attachment and neurite outgrowth assays. The results of cell attachment assays, presented in Fig. 10, show that neuroretinal cells' abilities to attach to these proteins is indeed regulated during development. While collagen IV remained an effective substrate for cell attachment from embryonic day 6 to 12, attachment to fibronectin declined with age rapidly and is lost by embryonic day 8. Attachment to laminin also declined with age and

was lost by embryonic day 10. Neurite outgrowth on laminin declined with approximately the same time course (Hall et al., 1987). The

164

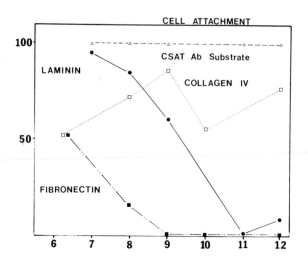

Figure 10. Developmental Change in Neural Retinal Cell Attachment to Laminin, Fibronectin, Collagen IV and Substrate-Attached CSAT Antibody. Neural retinal cells were obtained from embryos of different age as indicated. Cell attachment was measured on laminin (5 µg/ml; closed circles), fibronectin (300 µg/ml; closed squares), collagen IV (2 µg/ml; open squares) and CSAT antibody (50 µg/ml; open triangles). Percent of cells attached vs. embryonic day are shown. Each point is the average of at least three separate determinations.

integrin β_1 subunit remains on the surface of these cells at all ages, since all of the cells attached to substrate-bound anti-integrin β_1. This result suggests that α subunits of the integrin β_1 family are regulated differentially, and that α subunits that mediate binding to fibronectin and laminin are lost or inactivated in older neuroretinal cells.

To directly examine these possibilities, anti-integrin β_1 was used to immunoprecipitate metabolically labeled surface proteins from cultured chick neural retinal cells. Dissociated neural retinal cells were obtained from embryos of different ages, preplated to remove nonneuronal cells, and metabolically labeled with ^{35}S-methionine. Chick embryo fibroblasts were used for comparison to previously described proteins precipitated by anti-integrin β_1 (Horwitz et al, 1984; Hasegawa et al, 1985; Knudsen et al., 1985). Developmental differences in the retinal proteins were analyzed by nonreducing SDS-PAGE, where fibroblast glycoproteins were separated into four distinct components of Mr 145,000, 135,000, 125,000, 110,000 (Fig. 11). The Mr 110k glycoprotein has been shown

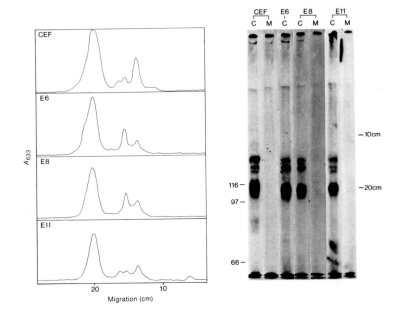

Figure 11. Nonreducing SDS-PAGE Analysis of Proteins Immunoprecipitated with Anti-Integrin β_1 Antibody from Chick Embryo Fibroblasts and Embryonic Neural Retinal Cells. (Right) Fluorograph of a 7.5% SDS-PAGE run under nonreducing conditions. Lanes marked C were loaded with CSAT immunoprecipitates from chick embryo fibroblasts (CEF) and embryonic neural retinal cells from E6, E8 and E11. Nonimmune mouse IgG precipitates are shown in adjacent lanes (M) for comparison. (Left) Densitometry of the autoradiograph. The chick embryo fibroblast (CEF) sample and neural retinal cell samples of different ages (E6, E8, or E11) are indicated.

to be the β_1 subunit (Horwitz et al., 1985). The other bands represent α subunits. The bands present in the retinal cell samples were similar, but not identical to those of fibroblasts. A band of Mr 145,000 was present both in chick embryo fibroblast and in retinal immunoprecipitates of each age. A middle band, of Mr 135,000, was present in all four samples, but migrated as a doublet in the fibroblast and E11 samples. Densitometry confirmed that this band migrated as a doublet. A diffuse protein band of Mr 110,000, presumably the β_1 subunit, was present in all four samples, and E6 retinal cells had an additional band of lower Mr. This band was less prominent in the E8 and E11 retinal samples and appears as a shoulder on the densitometric scan. Two dimensional gel analyses of identical samples has confirmed that there are many age related differences in the surface glycoproteins immunoprecipitated by anti-integrin β_1 (Neugebauer, unpublished observation). The nature of the relationship between these developmentally regulated differences in surface integrin subunits and responses to extracellular matrix molecules clearly warrants further investigation.

FURTHER STUDIES ON NEURORETINAL CELL INTERACTIONS WITH ECM AND CELLULAR SUBSTRATES.

Results described in previous sections suggest that the basal lamina that forms the inner limiting membrane is likely to be crucial for normal retinal ganglion cell axon growth. In addition, laminin seems likely to modulate retinal ganglion cell axon growth in the other places where it is found along the retinotectal pathway. It is now clear, however, that axon growth does not depend completely on interactions with ECM glycoproteins nor on the functions of integrin-class ECM receptors. Retinal ganglion cells of many ages have been shown to extend axons on astrocytes, a cell type with surface properties similar to those of the astroglia encountered by retinal ganglion cell axons in the retina, optic nerve and optic tectum (Cohen et al., 1986; Neugebauer et al., 1988). Neuroretinal axon extension on astrocytes at embryonic day 7 has been shown to depend on the function of integrin β_1-class ECM receptors, the cell-cell adhesion molecule N-cadherin, and other unidentified receptors (Neugebauer et al., 1988). At embryonic day 11, axon extension by the same neurons depends on two cell-cell adhesion molecules, N-cadherin and N-CAM, the integrin β_1-class of ECM receptors, and other molecules. As embryonic day 11 retinal ganglion cells have lost responsiveness to laminin and fibronectin (Cohen et al., 1986; 1987; Hall et al., 1987), the continued dependence on integrin β_1 function of axon outgrowth by these older neurons implies that astrocytes secrete additional ECM glycoproteins that promote neurite extension. These findings also suggest that the functions of cell adhesion molecules and integrins are both regulated during development. A future challenge will be to understand the factors that regulate integrin and cell adhesion molecule function on these neurons.

ACKNOWLEDGEMENTS

Work from the authors' laboratory was supported by NIH grant NS19090. DEH was a postdoctoral fellow of the Muscular Dystrophy Association; KMN is supported by a predoctoral fellowship from the National Science Foundation; LFR is an investigator of the Howard Hughes Medical Institute. Figures 3 and 10 contain data previously presented in Hall et al., 1987 and Figure 11 is reproduced from the same source by copyright permission of the Rockefeller University Press. We thank A.F. Horwitz for the CSAT monoclonal secreting hybridoma and Clayton Buck for polyclonal CSAT (anti-β_1) antibody. We thank Kevin Tomaselli and John Bixby for helpful discussions and Marion Meyerson for expertly typing this manuscript.

REFERENCES

Adler, R., Jerdan, J. and Hewitt, A.T. (1985) Responses of cultured neural retinal cells to substratum-bound laminin and other extracellular matrix molecules. Devel. Biol. 112:100-114.

Aumailley, M., Nurcombe, V., Edgar, D., Paulsson, M. and Timpl, R. (1987) The cellular interactions of laminin fragments. J. Biol. Chem. 262:11532-11538.

Bixby, J.L. and Reichardt, L.F. (1987) Effects of antibodies to neural cell adhesion molecule (N-CAM) on the differentiation of neuromuscular contacts between ciliary ganglion neurons and myotubes in vitro. Devel. Biol. 119:363-372.

Bixby, J.L., Lilien, J., and Reichardt, L.F. (1988) Identification of the major proteins that promote neuronal process outgrowth on Schwann cells in vitro. J. Cell Biol. 107, in press.

Bonhoeffer, F. and Huf, J. (1980) Recognition of cell types by axonal growth cones in vitro. Nature 288:162-164.

Bozyczko, D. and Horwitz, A.F. (1986) The participation of a putative cell surface receptor for laminin and fibronectin in peripheral neurite extension. J. Neurosci. 6:1241-1251.

Bray, D. and Chapman, K. (1985) Analysis of microspike movements on the neuronal growth cone. J. Neurosci. 5:3204-3213.

Buck, C.A., Shea, E., Duggan, K. and Horwitz, A.F. (1986) Integrin (the CSAT antigen): Functionality requires oligomeric integrity. J. Cell Biol. 103:2421-2428.

Chang, S., Rathjen, F.G. and Raper, J.A. (1987) Extension of neurites on axons is impaired by antibodies against specific neural cell surface glycoproteins. J. Cell Biol. 104:355-362.

Charonis, A.S., Tsilibary, E.C., Yurchenco, P.D. and Furthmayr, H. (1985) Binding of laminin to type IV collagen: A morphological study. J. Cell Biol. 100:1848-1853.

Chiu, A.Y., Matthew, W.D. and Patterson, P.H. (1986) A monoclonal antibody that blocks the activity of a neurite regeneration-promoting factor: Studies on the binding site and its localization in vivo. J. Cell Biol. 103:1383-1398.

Cohen, J., Burne, J.F., Winter, J. and Bartlett, P. (1986) Retinal ganglion cells lose response to laminin with maturation. Nature 322:465-467.

Cohen, J., Burne, J.F., McKinlay, C. and Winter, J. (1987) The role of laminin and the laminin/fibronectin receptor complex in the outgrowth of retinal ganglion cell axons. Devel. Biol. 122:407-418.

DeGeorge, J.J., Slepecky, N. and Carbonetto, S. (1985) Concanavalin A stimulates neuron-substratum adhesion and nerve fiber outgrowth in culture. Devel. Biol. 111:335-351.

Easter, S.S., Bratton, B. and Scherer, S.S. (1984) Growth-related order of the retinal fiber layer in goldfish. J. Neurosci. 4:2173-2190.

Edgar, D., Timpl, R., and Thoenen, H. (1984) The heparin-binding domain of laminin is responsible for its effects on neurite outgrowth and neuronal survival. EMBO J. 3:1463-1468.

Engvall, E., Davis, G.E., Dickerson, K., Ruoslahti, E., Varon, S. and Manthorpe, M. (1987) Mapping of domains in human laminin using monoclonal antibodies: localization of the neurite-promoting site. J. Cell Biol. 103:2457-2466.

Fallon, J.R. (1985) Preferential outgrowth of central nervous system neurites on astrocytes and Schwann cells as compared with nonglial cells. J. Cell Biol. 100:198-207.

Goldberg, D.J. and Burmeister, D.W. (1986) Stages in axon formation: observations of growth of Aplysia axons in culture using video-enhanced contrast-differential interference contrast microscopy. J. Cell Biol. 103:1921-1931.

Goodman, S.L., Deutzmann, R. and von der Mark, K. (1987) Two distinct cell-binding domains in laminin can independently promote nonneuronal cell adhesion and spreading. J. Cell Biol. 105:589-595.

Graf, J., Iwamoto, Y., Sasaki, M., Martin, G.R., Kleinman, H.K., Robey, F.A. and Yamada, Y. (1987) Identification of an amino acid sequence in laminin mediating cell attachment, chemotaxis, and receptor binding. Cell 48:989-996.

Gunderson, R.W. (1987) Response of sensory neurites and growth cones to patterned substrata of laminin and fibronectin in vitro. Devel. Biol. 121:423-431.

Halfter, W. and Chen, S.F. (1987) Immunohistochemical localization of laminin, neural cell adhesion molecule, collagen type IV and T-61 antigen in the embryonic retina of the Japanese quail by in vivo injection of antibodies. Cell and Tissue Res. 249:487-496.

Halfter, W. and Deiss, S. (1984) Axon growth in embryonic chick and quail retinal whole mounts in vitro. Devel. Biol. 102:344-355.

Halfter, W. and Deiss, S. (1986) Axonal pathfinding in organ-cultured embryonic avian retina. Devel. Biol. 114:296-310.

Halfter, W., Reckhaus, W., and Kroger, S. (1987) Nondirected axonal growth on basal lamina from avian embryonic neural retina. J. Neurosci. 7:3712-3722.

Hall, D.E., Neugebauer, K.M. and Reichardt, L.F. (1987) Embryonic neural retinal cell response to extracellular matrix proteins: Developmental changes and effects of the cell substratum attachment antibody (CSAT). J. Cell Biol. 103:623-634.

Hall, D.E., Bixby, J.L. and Reichardt, L.F. (1988) Central and peripheral neurons respond to different functional domains of laminin. In preparation.

Hasegawa, T., Hasegawa, E., Chen, W.-T., and Yamada, K.M. (1985) Characterization of a membrane-associated complex implicated in cell adhesion to fibronectin. J. Cell Biochem. 28:307-318.

Henke-Fahle, S. and Bonhoeffer, F. (1983) Inhibition of axonal growth by a monoclonal antibody. Nature 303:65-67.

Hoffman, S., Sorkin, B.C., White, P.C., Brackenbury, R., Mailhammer, R., Rutishauser, U., Cunningham, B.A. and Edelman, G.M. (1982) Chemical characterization of a neural cell adhesion molecule purified from embryonic brain membranes. J. Biol. Chem. 257:7720-7729.

Horwitz, A., Duggan, K., Greggs, R., Decker, C. and Buck, C. (1985) The cell substratum attachment (CSAT) antigen has properties of a receptor for laminin and fibronectin. J. Cell Biol. 101:2134-2144.

Jerdan, J., Lindsey, J.D., Adler, R. and Hewitt, A.T. (1984) Laminin immunoreactivity and extracellular spaces in chick embryo neural retina. Soc. Neurosci. Abstr. 10:40.

Kapfhammer, J,P., and Raper, J.A. (1987) Interactions between growth cones and neurites growing from different neural tissues in culture. J. Neurosci. 7:1595-1600.

Knudsen, K.S., Horwitz, A.F. and Buck, C.A. (1985) A monoclonal antibody identifies a glycoprotein complex involved in cell substratum adhesion. Exp. Cell Res. 157:218-226.

Krayanek, S. and Goldberg, S. (1981) Oriented extracellular channels and axonal guidance in the embryonic chick retina. Devel. Biol. 84:41-50.

Lander, A.D., Tomaselli, K.J., Calof, A.L. and Reichardt, L.F. (1983) Studies on extracellular matrix components that promote neurite outgrowth. Cold Spring Harbor Quant. Biol. 48:611-623.

Manthorpe, M., Engvall, E., Ruoslahti, E., Longo, F.M., Davis, G.E. and Varon, S. (1983) Laminin promotes neuritic regeneration from cultured peripheral and central neurons. J. Cell Biol. 97:1882-1890.

McCaffery, C.A., Raju, T.R. and Bennett, M.R. (1984) Effects of cultured astroglia on the survival of neonatal rat retinal ganglion cells in vitro. Devel. Biol. 104:441-448.

McLoon, S.C. (1984) Distribution of laminin in the developing visual system of the chick. Soc. Neurosci. Abstr. 10:466.

Mohon, P.S. and Spiro, R.G. (1986) Macromolecular organization of basement membranes. J. Biol. Chem. 261:4328-4336.

Neugebauer, K.M., Tomaselli, K.J., Lilien, J. and Reichardt, L.F. (1988) N-cadherin, NCAM and integrins promote retinal neurite outgrowth on astrocytes in vitro. Submitted.

Pytela, R., Pierschbacher, M.D. and Ruoslahti, E. (1985a) Identification and isolation of a 140 kd cell surface glycoprotein with properties expected of a fibronectin receptor. Cell 40:191-198.

Pytela, R., Pierschbacher, M.D. and Ruoslahti, E. (1985b) A 125/115 kDa cell surface receptor specific for vitronectin interacts with the arginine-glycine-aspartic acid adhesion sequence derived from fibronectin. Proc. Natl. Acad. Sci. USA 82:5766-5770.

Pytela, R., Pierschbacher, M.D., Ginsberg, M.H., Plow, E.F. and Ruoslahti, E. (1986) Platelet membrane glycoprotein IIb/IIIa: member of a family of Arg-Gly-Asp specific adhesion receptors. Science 231:1559-1562.

Ott, U., Odermatt, E., Engel, J., Furthmayr, H., and Timpl, R. (1982) Protease resistance and conformation of laminin. Eur. J. Biochem. 123:63-72.

Rager, G. (1980) Development of the retinotectal projection in the chicken. Adv. Anat. Embryol. Cell Biol. 63:1-92.

Rao, C.N., Margulies, I.M.K., Tralka, T.S., Terranova, V.P., Madri, J.A. and Liotta, L.A. (1982) Isolation of a subunit of laminin and its role in molecular structure and tumor cell attachment. J. Biol. Chem. 257:9740-9744.

Rogers, S.L., Letourneau, P.C., Palm, S.L., McCarthy, J. and Furcht, L.T. (1983) Neurite extension by peripheral and central nervous system neurons in response to substratum bound fibronectin and laminin. J. Cell Biol. 98:212-220.

Ruoslahti, E. and Pierschbacher, M.D. (1987) New perspectives in cell adhesion: RGD and integrins. Science 238:491-497.

Silver, J. (1984) Studies on the factors that govern directionality of axonal growth in the embryonic optic nerve and at the chiasm of mice. J. Comp. Neurol. 223:238-251.

Silver, J. and Sapiro, J. (1981) Axonal guidance during the development of the optic nerve: role of pigmented epithelia and other extrinsic factors. J. Comp. Neurol. 202:[521-538.

Smallheiser, N.R., Crain, S.M. and Reid, L.M. (1984) Laminin as a substrate for retinal axons in vitro. Devel. Brain Res. 12:136-140.

Terranova, V.P., Rao, C.N., Kalebic, T., Margulies, I.M. and Liotta, L.A. (1983) Laminin receptor on human breast carcinoma cells. Proc. Natl. Acad. Sci. USA 80:444-448.

Timpl, R., Rohde, H., Gehron-Robey, P., Rennard, S.I., Foidart, J.-M., and Martin, G.R. (1979) Laminin - a glycoprotein from basement membranes. J. Biol. Chem. 254:9933-9937.

Timpl, R., Johansson, S., van Delden, V., Oberbaumer, I. and Hook, M. (1983) Characterization of protease-resistant fragments of laminin mediating attachment and spreading of rat hepatocytes. J. Biol. Chem. 258:8922-8927.

Tomaselli, K.J., Reichardt, L.F. and Bixby, J.L. (1986) Distinct molecular interactions mediate neuronal process outgrowth on non-neuronal cell surfaces and extracellular matrices. J. Cell Biol. 103:2659-2672.

Tomaselli, K.J., Neugebauer, K.M., Bixby, J.L., Lilien, J. and Reichardt, L.F. (1988) N-cadherin and integrins: Two receptor systems that mediate neuronal process outgrowth on astrocytes. Neuron. 1:33-43.

Index